Production-Integrated Environmental Protection and Waste Management in the Chemical Industry

Claus Christ (Editor)

Production-Integrated Environmental Protection and Waste Management in the Chemical Industry

Claus Christ (Editor)

WILEY-VCH

Weinheim · New York · Chichester · Brisbane · Singapore · Toronto

Dr. Claus Christ
Kuckucksweg 10
65779 Kelkheim

> This book was carefully produced. Nevertheless, authors, editor, and publisher do not warrant the information contained therein to be free of errors. Readers are advised to keep in mind that statements, data, illustrations, procedural details or other items may inadvertently be inaccurate.

First Edition 1999

Library of Congress Card No.: applied for

A CIP catalogue record for this book is available from the British Library

Deutsche Bibliothek Cataloguing-in-Publication Data:

Production integrated environmental protection and waste manage : the article production integrated environmental protection / Claus Christ (ed.). - 1. Aufl. - Weinheim ; New York ; Chichester ; Brisbane ; Singapore ; Toronto : Wiley-VCH, 1999
 ISBN 3-527-28854-6

© WILEY-VCH Verlag GmbH, D-69469 Weinheim (Federal Republic of Germany), 1999

Printed an acid-free and chlorine-free paper.

All rights reserved (including those of translation in other languages). No part of this book may be reproduced in any form – by photoprinting, microfilm, or any other means – nor transmitted or translated into machine language without written permission from the publishers. Registered names, trademarks, etc. used in this book, even when not specifically marked as such, are not to be considered unprotected by law.

Composition and Printing: Rombach GmbH, D-79115 Freiburg
Bookbinding: Wilhelm Osswald & Co., D-67433 Neustadt/Weinstraße

Printed in the Federal Republic of Germany.

Lao-tzu said:
"What is esteemed on the Way is the capacity to change."

(Wen-tzu: Further teachings of Lao-tzu; first century B.C.E.)

Quoted in: Wen-tzu: "Further teachings of Lao-tzu, understanding the mysteries";
translated by T. Cleary, p. 102
Shambhala Publications, Inc.
Boston, Mass., 1991

Preface

The present trend in world trade is best described by the term "Globalization". The concept conveys the rise of the global corporate economy, i.e., the interconnection of the national economies with their individual goods and capital flows, caused by the removal of barriers to trade and companies' global orientation. This has led to an increased interdependence of production in various countries. Foreign investment is of great importance in order to be represented in important markets.

However a second dualistic element of globalization has become apparent. This is the total field in which all economic processes occur. To be exact: the environment in which and from which we live. This is described by "ecology". Thus we have an economic-ecological paradigm. However this does not mean that these two elements should lead to a win-lose outlook in the sense of a zero sum game. In fact we should strive for a mutual win-win outlook, i.e., a give-take strategy.

These is the position in which the chemical industry with its ability to produce products with innovative processes for the market finds itself. The best example of these innovative processes is the "production-integrated environmental protection", i.e., these processes take both economic and ecological factors into account. It is hoped that this book, with many examples from the work of the chemical industry will cast light on this field. The main aim of production-integrated environmental protection is the reduction and avoidance of waste. Thus integrated measures in production need to be carried out closely with waste management. This has meant that a chapter on this concept has been added to this book. This chapter is intended for industrial chemists, chemical engineers, those persons with management responsibilities in environmental protection and members of the research and academic professions who wish to integrate this subject into their professional activities.

Finally I can only thank all those without whom this book would have never been completed. This is intended initially for my colleagues, the authors who with numerous technical examples have helped bring this book to its present comprehensive state. Further I must thank the DECHEMA e.V. (German Society for Chemical Apparatus, Chemical Technology and Biotechnology) and the Hoechst Marion Roussel Deutschland GmbH, which greatly helped in the production of this book.

Claus Christ

List of Contributors

CLAUS CHRIST, formerly Hoechst Aktiengesellschaft, Frankfurt/Main, Federal Republic of Germany (Chap. 1, 2, 4, 5, Section 3.2.1)

JÜRGEN WIESNER, DECHEMA, Frankfurt/Main, Federal Republic of Germany (Section 3.1 and coordination of the original Chapter 3)

WOLFGANG FÜHRER, Bayer AG, Leverkusen, Federal Republic of Germany (Section 3.2.2)

HORST BEHRE, Bayer AG, Leverkusen, Federal Republic of Germany (Section 3.2.2)

HUBERTUS CUPPEN, Bayer AG, Leverkusen, Federal Republic of Germany (Section 3.2.2)

MICHAEL LUMM, Bayer AG, Leverkusen, Federal Republic of Germany (Section 3.2.2)

FRANZ-JOSEF MAIS, Bayer AG, Leverkusen, Federal Republic of Germany (Section 3.2.2)

GERHARD SCHROEDER, Bayer AG, Leverkusen, Federal Republic of Germany (Section 3.2.2)

FERDINAND SENGE, Bayer AG, Leverkusen, Federal Republic of Germany (Section 3.2.2)

DIETER STOCKBURGER, BASF Aktiengesellschaft, Ludwigshafen, Federal Republic of Germany (Section 3.2.3)

LUDWIG SCHMIDHAMMER, Wacker-Chemie GmbH, Burghausen, Federal Republic of Germany (Section 3.2.4)

GREGOR LOHRENGEL, Hüls Aktiengesellschaft, Werksgruppe Herne, Herne, Federal Republic of Germany (Section 3.2.5.1)

LOTHAR KERKER, Hüls Aktiengesellschaft, Marl, Federal Republic of Germany (Section 3.2.5.2)

HANS REGNER, Hüls Aktiengesellschaft, Marl, Federal Republic of Germany (Section 3.2.5.2)

ULRICH ROTHE, Kronos International, Inc., Leverkusen, Federal Republic of Germany (Section 3.2.6)

VOLKMAR JORDAN, Henkel KGaA, Düsseldorf, Federal Republic of Germany (Section 3.2.7)

BERNHARD GUTSCHE, Henkel KGaA, Düsseldorf, Federal Republic of Germany (Section 3.2.7)

THOMAS GLARNER, F. Hoffmann-LaRoche AG, Basel, Switzerland (Section 3.2.8)

KONRAD STOLZENBERG, VFT AG, Castrop-Rauxel, Federal Republic of Germany (Section 3.2.9)

JÖRG TALBIERSKY, VFT AG, Castrop-Rauxel, Federal Republic of Germany (Section 3.2.9)

CORNELIS VAN OS, Shell Internationale Chemie Maatschappij B.V., The Hague, The Netherlands (Section 3.2.10)

CHRISTOPHER HIGMAN, Lurgi Öl-Gas-Chemie GmbH, Frankfurt/Main, Federal Republic of Germany (Section 3.2.11)

ROBERTO DE PIAGGI, De Nora Permelec S.p.A., Milan, Italy (Section 3.2.12)

MOTOHISA MIYACHI, Mitsubishi Chemical Corporation, Tokyo, Japan (Section 3.2.13)

FUMIHIKO ODA, Mitsubishi Chemical Corporation, Tokyo, Japan (Section 3.2.13)

JUN YONAMOTO, Mitsubishi Chemical Corporation, Tokyo, Japan (Section 3.2.13)

GÜNTHER SCHUMACHER, Boehringer Mannheim GmbH, Werk Penzberg, Penzberg, Federal Republic of Germany (Section 3.2.14)

Contents

1.	Introduction	1
2.	**Production-Integrated Environmental Protection in the Chemical Industry**	5
2.1.	**Chemical Industry and Sustainable Development**	5
2.2.	**Formation of Residues in Chemical Processes**	6
2.3.	**Environmental Concepts in the Chemical Industry**	9
2.3.1.	Review of the Environmental Concepts	9
2.3.2.	The Concept of Integrated Environmental Protection	10
2.3.3.	Environmental Protection in Research and Development	11
2.3.4.	Integrated and Additive Concepts of Environmental Protection	12
2.3.5.	Comparison of Integrated and Additive Environmental Protection	15
2.3.6.	Methods of material flow and cost management	21
2.4.	**Limitations of Production-Integrated Environmental Protection**	27
2.4.1.	Technical Limitations	28
2.4.2.	Economic Limitations	29
2.5.	**Effect of Production-Integrated Environmental Protection**	29
2.6.	**Costs of Integrated Measures**	31
3.	**Examples of Production-Integrated Environmental Protection in the Chemical Industry**	33
3.1.	**Introduction**	33
3.2.	**Selected Examples**	34
3.2.1.	Examples from Hoechst	34
3.2.1.1.	Recovery and Utilization of Residues in the Production of Viscose Staple Fiber	34
3.2.1.2.	Recovery of Methanol and Acetic Acid in Poly (Vinyl Alcohol) Production	40
3.2.1.3.	Acetylation without Contamination of Wastewater	40
3.2.1.4.	Reutilization Plant for Organohalogen Compounds	41
3.2.1.5.	Vacuum Technology for Closed Production Cycles	45
3.2.1.6.	Utilization of Exhaust Gases and Liquid Residues of Chlorination Processes for Production of Clean Hydrochloric Acid Hydrochloric Acid Hydrochloric Acid Hydroc	47
3.2.1.7.	Production of Neopentyl Glycol: Higher Yield by Internal Recycling	50
3.2.1.8.	Optimization of Ester Waxoil Production and Recovery of Auxiliary Products	51
3.2.1.9.	Biochemical Production of 7-Aminocephalosporanic Acid	53
3.2.1.10.	Production of	56
3.2.1.11.	Production of Theobromine	57
3.2.1.12.	Recovery of Organic Solvents	59
3.2.2.	Examples from Bayer	67
3.2.2.1.	Avoidance of Wastewater and Residues in the Production of H Acid (1-Amino-8-hydroxynaphthalene-3,6-disulfonic acid)	67
3.2.2.2.	High-Yield Production of Alkanesulfonates by Means of Membrane Technology	71
3.2.2.3.	Selective Chlorination of Toluene in the	73
3.2.2.4.	Production of Naphthalenedisulfonic Acid with Closed Recycling of Auxiliaries	75

XI

3.2.2.5.	Avoiding Residues in Dye Production by Using Membrane Processes	80	3.2.12.	Neutral Salt Splitting with the Use of Hydrogen Depolarized Anodes (HydrinaTechnology, Example from De Nora Permelec). 142
3.2.2.6.	Fuel Replacement in Sewage Sludge Combustion by Utilization of Chlorinated Hydrocarbon Side Products.	81	3.2.13.	Ultrapure Isopropanol Purification and Recycling System (Example from Mitsubishi Chemical) 151
3.2.3.	Examples from BASF	84	3.2.14.	Examples from Boehringer Mannheim 156
3.2.3.1.	Emission Reduction in Industrial Power Plants at Chemical Plant Sites by Means of Optimized Cogeneration.	84	3.2.14.1.	Biocatalytic Splitting of Penicillin . 156
			3.2.14.2.	Production of Diagnostic Reagents by Means of Genetic Engineering: Glucose-6-Phosphate Dehydrogenase and 158
3.2.3.2.	Closed-Cycle Wittig Reaction	90		
3.2.4.	Integrated Environmental Protection and Energy Saving in the Production of Vinyl Chloride (Example from Wacker Chemie). .	93	**4.**	**Waste Management in the Chemical Industry** 163
3.2.5.	Examples from Hüls	104	4.1.	**Introduction** 163
3.2.5.1.	Integrated Environmental Protection in Cumene Production .	104	4.2.	**Chemical Industry Wastes** 163
3.2.5.2.	Production of Acetylene by the Hüls Plasma Arc Process.	107	4.3.	**Waste-Management Concepts** . . 164
			4.3.1.	Residues and Wastes from Production 164
3.2.6.	Low-Residue Process for Titanium Dioxide Production (Example from Kronos International)	111	4.3.2.	Production-Oriented Management. 165
			4.3.3.	Product-Oriented Management. . . 166
3.2.7.	Reduction of Waste Production and Energy Consumption in the Production of Fatty Acid Methyl Esters (Example from Henkel) . . .	115	4.4.	**Disposal Measures** 167
			4.4.1.	Logistics 167
			4.4.2.	Waste Combustion 168
			4.4.3.	Landfill Disposal of Wastes 173
3.2.8.	Integrated Environmental Protection in the Production of Vitamins (Example from F. Hoffmann-La Roche).	122	4.4.4.	Asbestos Disposal 174
			4.5.	**Utilization of Product Wastes** . 174
			4.5.1.	Plastics Recycling. 174
			4.5.2.	Refrigerant Recycling 176
3.2.9.	Production of Pure Naphthalene without Residues—Replacement of Chemical Purification by Optimized Multiple Crystallization (Example from	128	4.5.3.	Recycling of Used Packaging Materials 177
			4.5.4.	Paint Recycling 178
			4.6.	**Results of Waste Management** 178
			5.	**Summary and Outlook** 183
3.2.10.	Improvements in the Polypropylene Production Process (Example from Shell).	132	**6.**	**References**. 185
				Index 197
3.2.11.	The Zero-Residue Refinery Using the Shell Gasification Process (Shell – Lurgi Example)	135		

1. Introduction

Population growth and the rising material expectations of the peoples of the industrialized countries coupled with the expansion of the market economies are causing a global increase in production and consumption. Thus the global environment – seen as the interaction between nature and mankind – is increasingly under threat. Natural resources necessary for the production of goods are taken from the environment.

On the other hand the environment receives the wastes or pollutants from production processes. It is put under stress by utilization and consumption of resources as well as the destruction of the products. Overloading of the supply and sink function of the environment influences its control function, i.e. the preservation of the ecological equilibrium is threatened. This is shown in Figure 1 [1], [2].

This group of problems forms the basis of the "Brundtland Report", published by the World Commission on Environment and Development in 1987. A key role in this report is played by the concept of Sustainable Development:

"Sustainable Development meets the needs of the present without compromising the ability of future generations to meet their own needs" [3].

This recognizes that ecological, economic, and social development are to be seen as part of an internal unity. This unity in the sense of an equilibrium of the three factors cannot be understood as a harmonic model. The three factors oppose each other in conflict [4]. It is clear from Figure 2 [5], that if this equilibrium is disturbed, it can lead to environmental damage, economic decline or social unrest.

The United Nations Conference for Environment and Development in 1992 accepted this model (Agenda 21) as one of the aims of the international organization. The freedom of action in environmental and development politics for nations worldwide are given in this agenda [6].

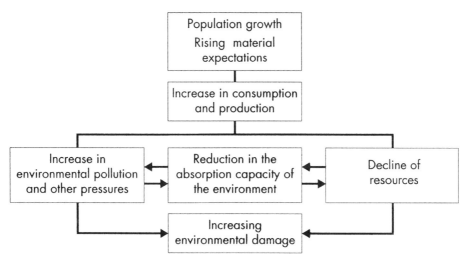

Figure 1. Factors threatennig the environment

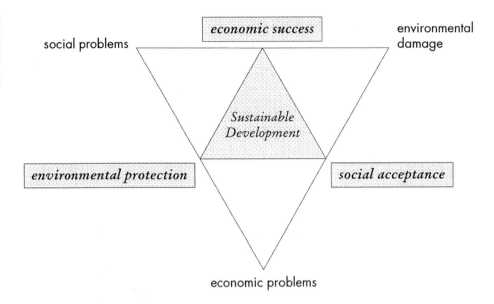

Figure 2. Factors of sustainability

Sustainability in an economic sense means an efficient allocation of scarce goods and resources. This presupposes a flexible economy with freedom of trade, i.e., free access to markets. This means that maximum innovation will be achieved, that changes will occur as efficiently and cost-effectively as possible and that the consumer will enjoy the best provision of goods and services under these conditions. Consequently unavoidable intervention by the state in the market processes should limit competition to the smallest extent [4].

Ecological sustainability means not exceeding the stress limits of the environment and preserving the basis of natural life. Social sustainability needs maximum equality of opportunity, freedom and social justice.

The aspect of justice has been newly formulated and extended in this respect by FABER [7]–[9] and MANNSTETTEN [10], so that it treats the question of a just distribution between the generations living today and future generations, i.e., those not yet born. Over and above this there is also the question of a fair treatment of the non-anthropoid natural world, i.e., the ecological factor. This borders on ethics [11]. To this end guidelines for the ethical assessment of companies on the basis of interdisciplinary cooperation were developed [12]. These guidelines contain amongst other aspects eight criteria under the aspect of compatibility with nature. Included in these are the development and application of innovations in the further development of environmental technology and the removal of the end-of-pipe technology. The authors of rating organizations (Standards and Poors, Moody, Eurorating) recommend the application of these models.

All in all it is necessary to distinguish between an operational way to sustainability and the ideal of sustainability [7]. Ideal sustainability can be compared to a star used by a sailor for navigation purposes. The ideal cannot be reached but it offers an orientation for the operator towards concrete sustainable development. Low-emission processes in conjunction with integrated environmental protection technologies are the tools for the operational way. Amongst these are also integrated production cycles. These processes are only viable when they are more cost effective in general than the application of conventional methods with bolt on separation and clean up systems [4], [13], [14].

2. Production-Integrated Environmental Protection in the Chemical Industry

2.1. Chemical Industry and Sustainable Development

The business of the chemical industry is to produce and sell chemical products. Profitability is therefore an essential aim of production. However, no chemical process exists that produces only the product desired. Other substances not desired by the producer are also formed in the gas, liquid, or solid state. These are referred to as residues.

Chemical production is thus a two-edged sword. On the one hand it manufactures products, i.e., "goods" and on the other hand it produces residues, i.e., "bads". The roman god "Janus" is here a good symbol (Fig. 3).

Another equally important aim is to reduce the environmental effects of these residues to a level that is acceptable. Therefore, protection of the environment — both as an objective in its own right and as a duty involved in any forward-looking activity — is a constituent part of the managerial policy of the German chemical industry [15]. The European chemical industry has also committed itself to a strategy of environmental protection within the program of Responsible Care. The chemical industry endeavors to operate the production process so as to minimize the entry of residues and waste products into the environment, both quantitatively and with respect to their hazard potential, and to utilize raw materials and energy with maximum efficiency. This growing awareness of the environment follows not only from the principle that the chemical industry is responsible for its actions, but also from the changed conditions governing the introduction of new production methods and the operation of existing plants. These include the following:

1) The demanding international and national decrees and regulations covering environmental protection
2) In chemical companies, the correct allocation of costs required for "cleaning up" a given production unit to comply with regulations (e.g., the purification of waste gas and wastewater and the disposal of waste materials from this unit)
3) Difficulties in the disposal of waste materials due to the shortage of dumping space and the secondary costs of waste disposal plants
4) Costs of raw materials and energy
5) Increased public awareness of the importance of environmental protection
6) In addition, the processes and production plants of chemical technology must meet high standards of workplace protection and operational safety.

Figure 3. Chemical production process

Thus, the chemical industry as a key industry can greatly contribute to the development of the concept of sustainable development.

It cannot offer a complete solution, but because of its knowledge and experience in handling substances, their processing and utilization as well as preparation and reuse it can mold important sections of this model [16], [17]. This is reflected in the following contributions:

– Improvement in value creation and productivity
– Optimal management of the raw materials and energy
– Environmentally friendly technologies as process improvements — as part of environmental protection and material cycles
– Technical processes allowing, e.g., safer disposal of waste

2.2. Formation of Residues in Chemical Processes

Chemical reactions do not lead exclusively to the desired product but also to residues which may pollute the environment. To understand the reasons for the formation of residues in chemical production processes, and to understand the possibilities for modifying these processes, the general conditions for chemical reactions must be considered. As an example an equilibrium reaction is considered in which a starting material A reacts with a reaction partner B to give a product P:

$$A + B + N \xrightarrow{M, C, H, E} P + P' + P'' + N'$$

where A is the starting material, B the reaction partner, N a secondary constituent of A and B, N′ the reacted secondary constituent, M the reaction medium, C a catalyst, H an auxiliary, P′ a joint product, E the energy, P the product, and P″ a side product or by-product.

The term "residues" denotes all components that take part in the reaction and do not give the desired target product P.

Incomplete Conversion. In equilibrium reactions, components A and B do not react with each other to completion. One of the two substances A or B is generally used in excess to increase the extent of conversion to the other at equilibrium. Occasionally, the component present in excess is the reaction medium for the chemical conversion. The excess amount of A or B remaining at the end of the reaction can contain a higher or lower content of impurities, and must be disposed of in an environmentally friendly manner if it cannot be recycled directly after physical or chemical treatment.

Associated Substances or Impurities in Raw Materials. The appearance of by-products or residues in chemical reactions can also be due to associated substances or impurities present in the starting materials. This applies especially to the winning of inorganic and organic basic materials from mineral, fossil, or biological raw materials. Examples of associated substances include the following:

1) Hydrogen sulfide in the processing of sour natural gases and mineral oils
2) Sulfur dioxide and dust emissions in the production of acetylene and synthesis gas by the formerly widely used coal-based processes
3) Contamination of wastewater by dissolved decomposition products of lignin (lignosulfonic acids, etc.) in the production of pulp from wood or other vegetable raw materials

Joint Products. Another characteristic of many syntheses is the formation of joint products P′, the quantity of which is coupled to the quantity of the target product P.

In inorganic syntheses, for example, the type of joint product can depend on the form of the raw material component A in the available mineral or ore. For example, when sulfidic nonferrous ores are treated in pyrometallurgical processes, SO_2 is formed as the joint product, and the joint product $FeSO_4$ is formed during the digestion of ilmenite ($FeTiO_3$) in the production of the white pigment TiO_2 by the sulfate process. In the production of phosphoric acid from phosphate rock by the sulfuric acid process the calcium and fluorine content of apatite even leads to two joint products, namely, gypsum and hydrogen fluoride.

In organic syntheses, starting material A often contains a reactive group that is split off in the form of a joint product during the reaction. In many cases, halogens or sulfonic acid groups are eliminated, forming acids or salts, and are replaced by other substituents. Also, selective reduction of organic compounds in aqueous solution by metals such as iron or zinc leads to formation of the corresponding metal salts or oxides as joint products.

By-products or Products of Secondary Reactions. By-products P" also reduce the yield of product and must be removed during the course of reaction or during processing of the reaction mixture. They can be formed by continuance of the reaction; e.g., in chlorination processes intended to produce monochlorinated compounds, higher chlorinated products can also be formed.

Reaction Medium. In chemical reactions and separation processes in the liquid phase (e.g., organic solvents or water), solutions or multiphase systems with at least one liquid component are present. The reaction medium or process medium M in these processes is either a solvent, and therefore not a contributor to the reaction, or both a solvent and a reaction partner, and is sometimes a catalyst.

Concentrated mineral acids (e.g., sulfuric acid or oleum) are often used in excess for digestion of inorganic raw materials or for production of reactive intermediates from relatively unreactive organic starting products. That proportion of acid not required for product formation must be either regenerated or disposed of in an environmentally friendly manner. Both routes are costly. Reprocessing is particularly difficult if the acid must be diluted or partially neutralized, or if salts have to be added to the solution in order to recover the product.

When water is used as the reaction medium, wastewater is produced that is sometimes highly contaminated with organic or inorganic substances.

Catalysts can also contribute to the formation of residues. Heterogeneous catalysts must be replaced when their activity is exhausted. In the case of homogeneous catalysts, either the catalyst component is removed along with the used process medium, or an additional processing stage is required to separate the used catalyst from the product.

Auxiliaries can be necessary in some reactions. Like catalysts, these take part in the process without being used up in the chemical reaction. For example, multiphase reactions sometimes require the use of surface-active agents to increase the interfacial surface area or accelerate the transfer of matter. These substances appear as residues or in wastewater when the reaction mixture is processed.

Process Energy. The process energy (electricity, heat) required must also be included in the environmental assessment of chemical processes. Generation of this energy from primary energy carriers also leads to emissions.

2.3. Environmental Concepts in the Chemical Industry

2.3.1. Review of the Environmental Concepts

Environmental protection in the chemical industry is divided into product related and production related areas. Environmental protection related to products covers the development and production of environmentally friendly products (e.g., paints, herbicides/pesticides, washing powder) and treatment of product wastes from processing and consumption (Chap. 4). Environmental protection related to production covers the concept of the production-integrated environmental protection and additive environmental protection. Additiv environmental protection is the German term for end-of-pipe technology. This further subdivision can be examined in Figure 4 [18].

The third concept covers production plant including the subsidiary areas of storing and packing of raw materials and products. It is necessary to construct and operate these, so that material loss does not lead to environmental damage. Furthermore this includes retention systems for contaminated water draining out of burning plant

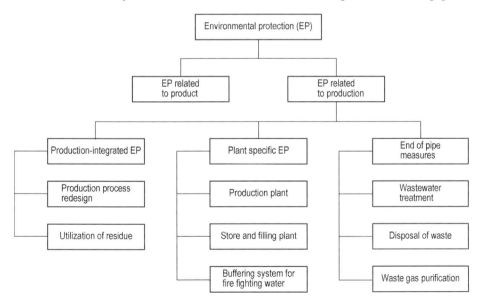

Figure 4. Environmental protection

during fires and containment systems for cooling water to prevent contamination of groundwater or surface waters.

2.3.2. The Concept of Integrated Environmental Protection

The idea of production-integrated environmental protection was implicitly described by A. W. von Hofmann (founder of the Royal College of Chemistry in London) as long ago as 1848: "In an ideal chemical factory ... there is, strictly speaking, no waste but only products (main and secondary by-products). The better a real factory makes use of its waste, the closer... it gets to its ideal, the bigger is the profit" [19].

In Germany, the concept of integrated environmental protection was introduced into the debate on environmental policy at the end of the 1970s [20]. It is being used to an increasing extent and is regarded as a mark of good environmental practice (e.g., in the report on the environment produced by the German Ministry of the Environment [21]). The term is defined differently in various social sectors. The following definitions are given as examples:

German Ministry for the Environment [22], [23]. Environmental protection must be integrated into all phases of the production cycle, i.e.,

1) Reduction or avoidance of harmful materials in the production process
2) Use of internal recycling and increase in energy efficiency in the production process
3) Development of alternative materials
4) Inclusion of the question of disposal in design of the product
5) Establishment of external recycling of production residues

Industrial Management [24]. The following measures should be put into effect:

1) Avoidance of end-of-the-pipe techniques where possible — that means process design that avoids or reduces residues
2) Control over the environmental effects of pre-stages and subsequent stages of the process
3) Cooperation with suppliers, consumers, and waste disposal agencies

Matter and energy balances for the process itself and its pre-stages and subsequent steps are taken into consideration when setting up these measures.

Institut für ökologische Wirtschaftsforschung (Institute for Ecological Economic Research), Berlin [25]. Integrated environmental protection techniques either do not cause environmental pollution compared with traditional techniques or do so to a lesser extent. Recycling is regarded as an integrated environmental protection method only if it is an integral part of the production process, i.e., if a closed-cycle operation is used that reduces emissions. If recycling occurs separately from the production process, it is regarded as an additive technique.

Statistisches Bundesamt (German Federal Bureau of Statistics) [26]. The term "integrated environmental protection" is not used in statistics. Instead, distinction is made according to those parts of fixed assets that are concerned with environmental protection. The corresponding costs of a process change are also included in the statistics.

European Union [27]. The term "integrated techniques" includes

1) Clean production techniques
2) Processes based on materials that result in low environmental pollution
3) Reprocessing and recycling systems

Other Related Terms [28], [29]. The terms "clean production processes," "clean production," and "clean technology" are used predominantly in English-speaking countries. *Clean production* goes to the roots of environmental problems, and minimizes both emissions and costs. *Clean technology* is more than a "set of clean production techniques;" rather it is an approach to provide users with services and other benefits enabling the environment to be protected from excessive pollution.

The following terms should also be mentioned: "best available technology" and "nonhazardous production." However, none of these terms have been clearly defined or distinguished from each other.

Verband der Chemischen Industrie (VCI, Association of the German Chemical Industry) [30]. In integrated environmental protection, the aim is to devise a process that causes as little pollution of air, water, and soil as possible, and in which integrated production methods enable residues to be utilized as far as possible. The technical and economic aims of the production process must nevertheless be fulfilled. This definition is also used by the Swiss chemical industry [31].

2.3.3. Environmental Protection in Research and Development

Chemical research is the precursor of production. Its aim is to devise solutions to problems that are valid for the long term. The total time required for research, process development, and planning and designing a plant (i.e., until the start of production) is at least six years. New developments must therefore be planned so that even after this time they are still state of the art. Moreover, they are future-oriented only if they take into account the requirements of environmental protection. This means that the environmental protection technology must also be future-oriented and gives extra impetus to developments in the field of environmental protection. Innovation in production methods is therefore an essential precondition for production-integrated environmental protection.

Improvements in environmental protection require capital investment. The introduction of new processes and new production plants always provides an opportunity for further improvement of environmental parameters. Capital investment must therefore be very carefully planned from an environmental point of view. This is ensured by a system of checklists that accompanies a new process from the research stage to the granting of a license by the authorities. At an early stage, in addition to substance data, environmental data are included in evaluation of the process; these are extended as the process is developed stage by stage [32], [33]. In this way, potential environmental problems can be recognized and dealt with at an early stage. The process must be optimized with respect to the amount of residues produced, which should be as low as possible. The possibility of using production methods that lead to smaller quantities of residues or less environmental pollution must therefore be investigated. The utilization of the Material Flow Analysis and Cost Flow Analysis offers the possibility of ascertaining how well ecological and economic aims have been reached. The result of this investigation can also be abandonment of a project.

Thus, production-based environmental measures are an integral part of research, development, and planning.

2.3.4. Integrated and Additive Concepts of Environmental Protection

The environmental demands placed on the production process can be met with the aid of integrated or additive (end-of-the-pipe [34]) concepts.

The sustainable development model demands a new orientation of environmental politics. Therefore innovations in all areas of industrial environmental protection are the most effective tools. These will lead to an efficiency revolution. Its elements can be characterized as follows:

- Process innovation: Production of the same or similar products using less raw material and energy and giving lower pollutant output as waste gas, waste and polluted wastewater, i.e., development of environmentally friendly processes.
- Material cycles: recycling of residues and product wastes can reduce use of resources.

There the adopted way of integrated environmental protection must be consistently followed further.

Production-Integrated Environmental Protection. According to the definition of the VCI, integrated environmental protection is production-linked, and that definition is used in this article. Thus, production-integrated environmental protection means measures taken to reduce, prevent, and utilize residues [31], [35]–[38].

The reduction and prevention of residues can be achieved by

1) Improving the chemical process with the aid of *new synthesis routes*. For example, in the production of aromatic amines, chemical reduction with iron chips is replaced by catalytic reduction by hydrogen.
2) *Shifting the equilibrium.* The use of more favorable reaction conditions can cause the position of the equilibrium to be shifted so that one of the two components A or B is almost 100 % reacted. This can be achieved by using the second component in excess, by removing the product, or by using more favorable temperature or pressure.
3) *Improving selectivity.* A very effective method of reducing the amounts of residues and improving the yield is to increase the selectivity of the chemical reaction. Examples of this include the following:
   – Improvement of the selectivity of catalysts, e.g., by using catalysts that lower the rate of an undesired side reaction
   – Maintenance of high catalytic activity, e.g., by avoiding contact poisons or by developing simple reaction methods
   – Optimization of reaction conditions, e.g., by utilizing differences in the reaction kinetics of the main reaction and the side reaction, more favorable temperature profiles and residence times, or more suitable reactors
   – Recycling of the side product (if the side reaction is reversible)
4) Developing new catalysts, e.g., in production of polypropylene without generation of wastewater using improved metal–organic catalysts.
5) Process optimization.
6) Changing the reaction medium. If water is replaced by an organic solvent in syntheses, contamination of wastewater can often be drastically reduced. However, environmentally friendly solvent handling involves not only recovery of the solvent from liquid media but also prevention of losses to the atmosphere during storage, transport, production, and subsequent processing. This can be achieved, for example, by adsorptive recovery of solvents from the gas stream.
7) Using raw materials of higher purity.
8) Replacing or eliminating auxiliaries that have a harmful effect on the environment (e.g., chlorinated hydrocarbons).

Gaseous, liquid, or solid residues whose formation during the production process cannot be avoided, even under optimum operation conditions, can often be reused (see. Fig. 5) by methods such as

1) *Internal utilization*, e.g., of the auxiliaries employed in the process, by processing and recycling directly into the process. In the simplest case, the substance can be used directly after separation from the products or process streams (recovery of volatile components of solvents). In other cases, a physical or chemical processing stage is necessary to remove impurities from the recycled components or convert them into a reusable form. Examples include

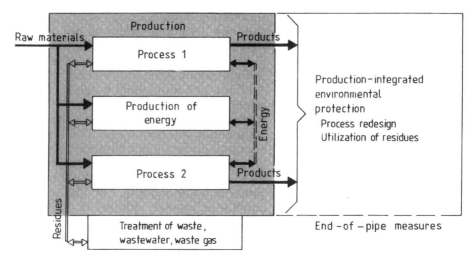

Figure 5. Approaches to environmental protection in chemical production

– Recovery of sulfuric acid from waste sulfuric acid by concentration or decomposition and reprocessing
– Recovery of organic solvents from solvent residues, solvent – product mixtures, aqueous solutions, and production residues
– Thermal decomposition of residues of chlorination processes to give pure hydrochloric acid

2) *external utilization* of the residue, i.e., as a raw material for the manufacture of other products in a separate production plant.

Reutilization of residues by linking of production processes is not another form of residue disposal, but enables resources to be used as economically as possible. However, this does lead to additional interdependencies, which may decrease flexibility.

Additive Environmental Protection. If the technical and economic possibilities for preventing, avoiding, or reusing residues (as part of a production-integrated environmental protection program) are exhausted, additive environmental technologies must be used:

1) Disposal of waste by land-filling and incineration
2) Techniques for purifying waste gas and wastewater

Of course, not all problems of environmental protection in the chemical industry can be solved by new process concepts based on the principles of prevention and utilization of residues. In many cases, additive processes are still necessary.

In the past, the main emphasis was on additive environmental protection employing conventional purification and disposal methods. Nevertheless, methods of recovery and utilization have long been used in the chemical industry, and can even be found at the beginning of industrial production [38].

In the chemical industry a long tradition exists of reprocessing reaction mixtures (e.g., by solvent extraction or distillation), with recovering of auxiliaries (e.g., organic solvents), and recycling these into the production process (internal recycling). For example, a desired product that is dissolved in a solvent following an extraction process can be recovered only by distilling off the extraction agent. This yields the solvent in a pure form that can be recycled, and such recycling is an inherent part of the production process. It is also a "prevention" since the residues remain in the production plant [39]. Changes in the production process aimed at increasing the yield or reducing the generation of residues have always been tasks of chemical research. Nevertheless, the technology of integrated environmental protection is becoming increasingly important today because current requirements with respect to the prevention and utilization of residues are much more stringent than ever before [40]–[45].

GLAUBER, an early exponent of technical chemistry, described in 1648 the conversion of wine yeast. Yeast is a by-product of wine production and its conversion lead at that time to new products coupled with a partial wastewater recycle (Figure 6). Pressing the yeast cake yielded wine for vinegar production. Distillation of the residues yielded spirits and further processing gave potash and tartar. Potash was supplied to the dyestuffs industry and tartar sold to hatters. The acid tartar-containing wastewaters were recycled in part. Some was used in treatment of copper ores to yield copper. GLAUBER also stressed the economic advantages of this process: Raw material costing 4 Thaler yielded an income of 16 Thaler from the products [46]. This is an early example of an integrated production method in utilization of a by-product and reduction of polluted wastewater with added benefit.

Product-Integrated Environmental Protection. The concept of *product*-integrated environmental protection should be distinguished from that of *production*-integrated environmental protection. The former includes the utilization of product waste (e.g., the recycling of plastics) and the development of environmentally friendly products. These measures are also part of an overall environmental protection strategy of the chemical industry [47]. They form a separate subject that is not discussed here.

2.3.5. Comparison of Integrated and Additive Environmental Protection

In additive environmental protection, harmful substances formed in a production process are disposed of at a later stage. There is no saving of raw material; instead, residues are converted into a comparatively environmentally harmless form by the use of additional raw materials and energy.

Additive techniques represent unproductive capital [48]. Moreover, these techniques do not reduce the consumption of raw materials. *Integrated techniques* reduce the consumption of raw materials and the production of waste [49]. However, new integrated techniques can lead to higher energy consumption. Integrated techniques some-

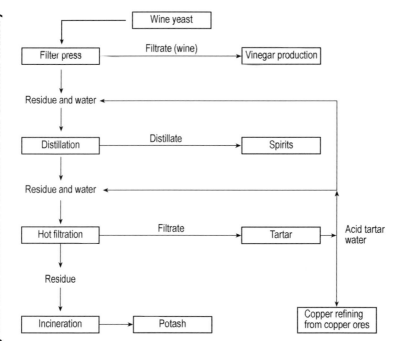

Figure 6. Conversion of wine yeast as an integrated production method

times offer economic advantages, whereas additive techniques with linear emission reduction lead to increased marginal costs or greatly decreased marginal utility [50]. This is shown for three German chemical companies taken as examples. Here, the use of additive environmental protection gave an increase in operating costs (Fig. 7), a reduction of emissions into the atmosphere (Fig. 8) and residual contamination of wastewater (Fig. 9). The extent of purification currently achieved in wastewater treatment plants cannot be improved significantly even at greater expense. The residual contamination of wastewater can be reduced only by techniques of integrated environmental protection. A similar development can also be seen in the case of waste from chemical production. The slight reduction in operating costs for environmental protection observed in 1994 and 1995 shows that integrated techniques are to some extent beginning to have an economic effect. However, if the operating costs of environmental protection are expressed as a percentage of value added, a further increase is observed (Fig. 10).

In contrast to additive technologies, the costs for operating and managing the new process are sometimes greater in the case of integrated techniques [51], [52]. Additive techniques are widely available and well proven or can be developed and installed in ca. two to four years. Integrated techniques, however, require research, development, and the approval of licensing authorities, which can take six to ten (or even more) years before they are incorporated into the production process [53], since the entire process

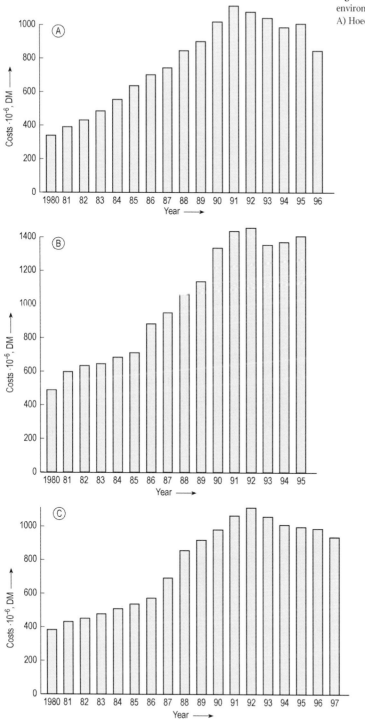

Figure 7. Operating costs for environmental protection
A) Hoechst; B) Bayer; C) BASF

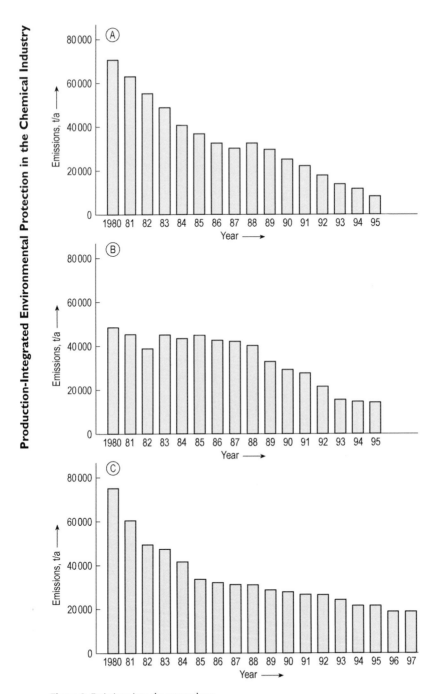

Figure 8. Emissions into the atmosphere
A) Hoechst; B) Bayer; C) BASF

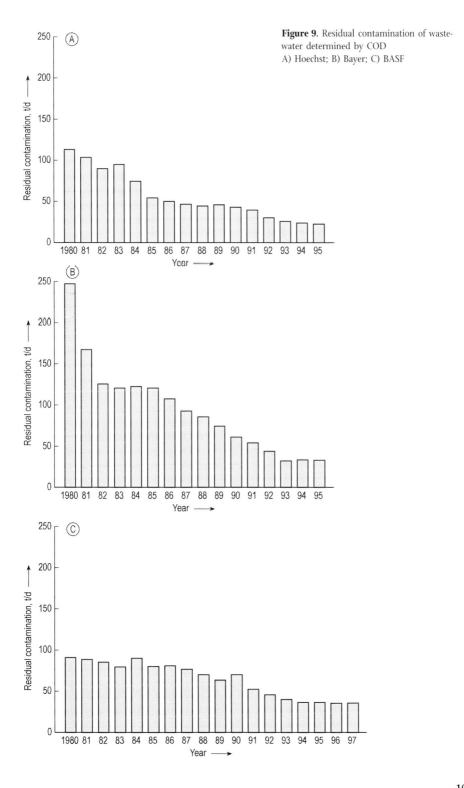

Figure 9. Residual contamination of wastewater determined by COD
A) Hoechst; B) Bayer; C) BASF

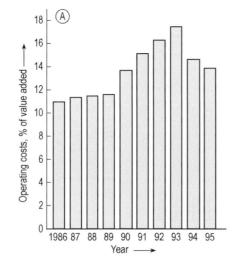
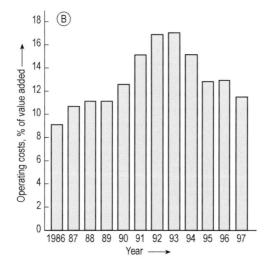

Figure 10. Operating costs of environmental protection in relation to added value
A) Hoechst; B) BASF

must often be designed afresh. Administrative barriers of environmental licensing authorities may also hinder the introduction of complex and innovative integrated techniques [48]. But the integrated environmental technology affects the end-of-pipe solution. The less polluted wastewater led to a reduction in loading of the biological wastewater treatment plant with respect to COD. The differences between end of pipe and integrated technology in both from economic and ecological viewpoints are summarized in Tables 1 and 2 using [54]. On the other hand, additive technologies for the disposal of materials, such as land-filling and waste incineration plants can meet resistance from the public.

As a result of organizational changes at Bayer and Hoechst, the data for the years following 1995 cannot be compared to the data for earlier years and have thus not been given here.

Table 1. Comparison of end-of-pipe and integrated environmental technology from the cost viewpoint

	End-of-pipe technology	Integrated technology
Total productivity	Reduction in productivity	Potential for increased productivity
Production costs	Increasing	Potential for cost reduction
Capital investment requirement	Lower	Higher
Depreciation of production ("sunk costs")	Generally none	Possible
Research and development costs	Lower	Higher
Adaption and conversion costs	Lower	Higher
Operating compatibility	Higher	Lower
Commercial risk	Lower	Higher
International market position (in environmental technology)	At present very good	Potential for a very good position
International competitiveness (of the overall economy)	Tends to reduce	Potential for future competitive advantages

Table 2. Comparison of end-of-pipe and integrated environmental protection technology from the ecological viewpoint

	End-of-pipe technology	Integrated technology
Energy and material efficiency	Lower	Higher
Potential for reducing pollution	Specific pollutants	Broader spectrum of pollutants
Temporal postponement or transfers to different medium	High	Lower but not ruled out
Potential for solving environmental problems	Not for all environmental problems	Potential for solving all environmental problems
Compensation for pollution-reducing effects	Possible	Possible

2.3.6. Methods of material flow and cost management

The aim of production is increasing productivity, i.e., economic effectiveness of raw material utilization, and lowering costs.

Stoichiometric Yield. material planing productivity is normally calculated in terms of the stoichiometric yield. This is based on the reaction equation, which describes the chemical process in question in the form of an ideal model. It allows the calculation of the theoretical amount of the target product given the amount of the main educt chosen. The stoichiometric yield of the target product is the ratio of the actual amount produces to the theoretical amount expected [55]. The stoichiometric yield is calculated on one educt and is thus dependent on the substance chosen as the main educt. The stoichiometric yield will therefore be designated the relative yield RY. Thus:

$$RY = \frac{\frac{m_{TP}}{n_{TP} \cdot M_{TP}}}{\frac{m_{ME}}{n_{ME} \cdot M_{ME}}} = \frac{m_{TP} \cdot n_{ME} \cdot M_{ME}}{m_{ME} \cdot n_{TP} \cdot M_{TP}}$$

m_{TP}	Real quantity of target product
m_{ME}	Real quantity of main educt
M_{TP}	Molecular weight of target product
M_{ME}	Molecular weight of main educt
n_{TP}	Number of moles of target product from reaction equation
n_{ME}	Number of moles of main educt from reaction equation

If the number of moles n of main educt and target product based on reaction equation are equal, then $n_{TP} = n_{ME}$ and

$$RY = \frac{m_{TP} \cdot M_{ME}}{m_{ME} \cdot M_{TP}}$$

The stoichiometric or relative yield has the following disadvantages: It yields no information on:

– other raw materials whether in excess or not
– the amounts of joint products
– auxiliary materials present, e.g., solvents

It is thus an "actual/theoretical relationship" and not an "input/output relationship". It is thus not a productivity parameter.

Material Flow Analysis (MFA). It is possible however to estimate a productivity parameter based on the reaction equation (ideal model). This parameter is the theoretical balance yield BA_t defined as follows:

$$BA_t = \frac{\text{Moles target product}}{\text{Moles of primary raw materials}} = \frac{(nM)_{TP}}{\sum (nM)_{PRM}}$$

Primary raw materials: All raw materials in the reaction equation.

This parameter is equivalent to "atom utilization" suggested by SHELDON [56]: "However, one category of selectivity is largely ignored by organic chemists: what I call the atom selectivity or atom utilization. The complete disregard of this important parameter is the root cause of the waste problem in fine chemical manufacture." The parameter is a constant of the synthesis route and is a measure of material utilization. As BA_t increases, the smaller the amount of joint products theoretically produced in the synthesis route chosen. BA_t is thus the maximum possible value of material planning productivity, because it is derived from the theoretical reaction scheme of the process, i.e., based on the stoichiometric reaction equation.

Material balances of chemical processes which cover all input and output substances, as well as residues, are necessary to determine the actual values of productivity. BA, the real balance yield under realistic conditions is defined as:

$$BA = \frac{\text{Amount target product}}{\text{Amount primary and secondary raw materials}} = \frac{m_{TP}}{\sum m_{PRM} + \sum m_{SRM}}$$

Secondary raw materials are substances which are present, but which do not take part in the reaction (according to the reaction equation) to form the target product. They can take part in subsidiary reactions. Typical secondary raw materials are solvents and catalysts.

BA (real balance yield) and BA_t (theoretical balance yield) play a central role in material flow analysis MFA (Trade Mark of BTC Dr. Dr. Steinbach GmbH, Mannheim, Germany).

The ratio of these two values

$$spBA = \frac{BA}{BA_t}$$

shows how near the actual productivity approaches the theoretical value. spBA, the specific balance yield is a measure of the degree of optimization of a process. The productivity function

$$BA = BA_t \cdot spBA$$

demonstrates both parameters of productivity, i.e., BA_t process quality and spBA the degree of optimization.

The use of these parameters RY, BA and BA_t was illustrated utilizing the production of phenone via the Friedel–Craft acetylation. The basic balance sheet (material balance) is shown in Figure 11. The values of BA and BA_t were calculated from the complete reaction equation (Figure 12). ($AlCl_3$ and 3 H_2O are included as primary raw materials.) The following value was determined for the stoichiometric or relative yield (calculated on benzoyl chloride as main educt):

$$RY = \frac{1000 \cdot 140.6}{700 \cdot 210.2} = 95\%$$

Comparing the above value with the actual values determined:

$BA_t = 48\%$, $BA = 30\%$, $RY = 95\%$

it is clear that the stoichiometric yield does not reflect the process productivity, but in fact is misleading. The degree of optimization is:

$$spBA = \frac{30}{48} = 62.5\%$$

	Input		Output		
Primary raw materials	Benzoyl chloride	700	Phenone	1000	Target product
	o-Xylene	550			
	Aluminum chloride	700	Aluminum chloride	410	Wastewater
	Water for reaction	258	Hydrochloric acid	600	
			other	123	
Secondary raw materials	Toluene	900	Toluene	900	Incineration
	Sulfuric acid	192	other	267	
	Balance sheet total	3300	Balance sheet total	3300	

Figure 11. Basic balance sheet: phenone in kg (simplified)

This means that the process optimization has reached 62.5% of the theoretically possible productivity. The degree of optimization can be interpreted taking the various influencing factors into consideration:

The material flow model of the MFA differentiates between two types of material flow:

– Main reaction stream: The main reaction, and this alone, leads to the target product. The input to the main reaction stream is made up of those primary raw materials found in the main reaction equation. A part of these raw materials can also be found, changed or unchanged, in the residues, i.e., as excess amounts of raw materials.

– Other raw materials stream: The input of these raw materials forms the secondary raw materials. These all end up as residues. The secondary raw materials include the non-recyclable solvents, catalysts and impurities (side components) in the primary raw materials.

These two types of material stream should be considrered and reviewed separately in process optimization. These approaches lead to the detailed productivity function [57], [58]:

$$BA = BA_t \cdot (f \cdot RY) \cdot EA_p$$

where

$$spBA = f \cdot RY \cdot EA_p$$

BA   real balance yield (in MFA)
BA_t theoretical balance yield (in MFA)
RY   relative or stoichiometric yield
EA_p proportion of the main reaction stream of the total input

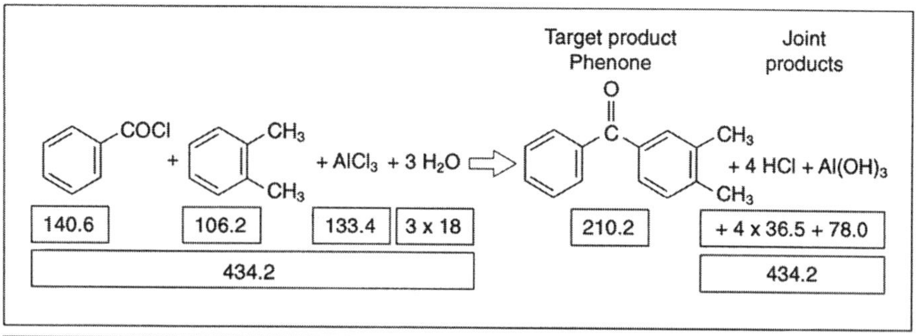

```
                                          Target product        Joint
                                          Phenone              products
                                             O
                                             ‖
   ⌬—COCl    ⌬—CH₃                          ⌬—C—⌬—CH₃
         +       |   + AlCl₃ + 3 H₂O  ⟹          |        + 4 HCl + Al(OH)₃
               CH₃                              CH₃
    140.6      106.2     133.4    3 × 18       210.2       + 4 × 36.5 + 78.0
              434.2                                        434.2
```

⇨ **Theoretical balance yield (values from stoichiometry)**

$$BA_t = \frac{\text{Output target product}}{\text{Input raw materials}} = \frac{210.2}{140.6 + 106.2 + 133.4 + 3 \times 18} = 48\%$$

⇨ **Real balance yield (values from MFA basic balance sheet)**

$$BA = \frac{\text{Real quantity of target product}}{\text{Real quantity of raw materials}} = \frac{1{,}000}{3{,}300} = 30\%$$

Figure 12. Balance yields BA_t and BA

$$EAp = \frac{\text{Amount of primary raw materials}}{\text{Amount of primary and secondary raw materials}} = \frac{\sum m_{PRM}}{\sum m_{PRM} + \sum m_{SRM}}$$

f = factor taking into account parameters ignored by the main reaction stream, e.g., excess amounts of primary raw materials.

The first four values are calculated by MFA material balances. The factor f can be calculated using the productivity function.

The limited value of RY, the stoichiometric yield is illustrated by the detailed productivity function. It offers only one aspect of BA, the productivity. To be more exact, RY describes only one part of the productivity of the main reaction. To give an example, if RY is increased by an excess amount of raw materials, this is counter-productive in the exact sense of the word: f is decreased. This will not be discussed further as a mathematical demonstration would go into too much detail here. The productivity function shows two furter weaknesses of RY, stoichiometric yield. This takes neither the importance of synthesis design nor the other raw materials (i.e. secondary raw materials and side components of the primary raw materials) for the productivity (or costing) into account.

The value of the stoichiometric yield of a process is of secondary importance for the productivity and thus the production-integrated environmental protection. The material planning quality BA_t of a process is more important, because the aim of production-integrated environmental protection is an increase in the material planning productivity [58a]. MFA as a computer based method allows systematic management via parametic studies. The main aim is the maximization of the output of the target

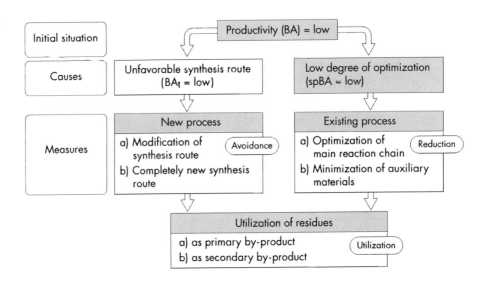

Figure 13. System for production-integrated environmental protection by MFA

product, i.e., BA → max. This systematic representation is derived from the productivity function (Figure 13).

- In case BA_t of the chosen process route is low, then a new process must be developed (reduction/avoidance residues).
- In case degree of optimization (spBA) is low, there are two alternatives (reduction of residues): either optimization of the main reaction stream, e.g., by reducing or recycling excess amounts of the primary raw materials and/or optimization of the other raw materials, e.g., by recycling solvents.
- The secondary aim is to make use of residues as primary or secondary by-product. Primary by-products are residues that are also generated in a definite production process besides the target product and economically used without prior treatment, i.e. raw materials for other syntheses. Secondary by-products are the resulting residues from a process that are treated outside the considered unit in order to gain immediately usable value-added products. The outside treatment also mostly generates waste materials. Secondary by-products are mixtures. The main components are considered those mixture components that represent the usable portion of the material.

MFA is also a method of carrying out material flow analysis and ecobalances.

Cost Flow Analysis (KFA). In addition to the material planning productivity cost is also an important factor. KFA is an integrated cost calculation, in which the chemical process, not the product, is most important. KFA utilizes the structure of the process from MFA and evaluates this with prices. This yields the process costs. Theses are made up of the raw material costs, waste costs and the manufacturing costs. The full manufacturing costs are then the calculated costs apportioned to the target product from the process costs. The KFA (Trade Mark of BTC Dr. Dr. Steinbach GmbH, Mannheim, Germany) computer program also allows determination of the environmental protection costs (proportion of the process costs arising from environmental aspects) and target costing [59]. This is based on the specified quality and the market price. The "target costs", i.e., the maximum allowable cost is then derived. Thus KFA shows how far the process optimization to increase the productivity lies at any time from the cost target and identifies areas of deviation. In production-integrated measures other factors: "profit of the chemical process" and "return on investment (ROI)" are of importance. These can be determined with Value Flow Analysis (WFA). This like MFA is based on KFA.

Thus MFA, KFA and WFA form a practical, scientific and technically correct basis for material planning, ecological and economic decision and control systems for increasing resource utilization productivity and improving ecological and economic efficiency.

Other Tools. Another method of estimating the industrial environmental pollution is the MIPS system (Material Intensity per Service Unit) [60]. This system is not suitable for the chemical industry corresponding to its claim "as assistance for the planning of environmentally friendly processes" [60], because it is not commercially practical. The application of MIPS uses consideration of life cycle as a factor for the general economy [61]. This application should lead to a non-material or dematerialized understanding of need satisfaction. In addition this system implies an increased devolution from property to utilization rights (as a new meaning of ownership), a change in consumption from material to the non-material goods going as far as a change in lifestyle coupled with the distribution questions associated with affluence [60], [62].

Another system is being developed at present under the abbreviation MFA, i.e., "Material Flow Accounting" [63], [64]. This utilizes a more general economic approach than MFA, Material Flow Analysis.

2.4. Limitations of Production-Integrated Environmental Protection

When integrating environmental protection into production processes, three main questions need to be asked for practical purposes [37]:
- what is technically feasible?
- what is ecologically worthwhile?
- what is economically feasible?

Processes must always be reviewed not only from an ecological standpoint but also from other points of view to avoid achieving only partial improvements. Questions of safety, material properties, energy and also economics, such as the situation of the world market prices, must be dealt with and the various factors objectively weighed against one another. Provided that sufficient material and energy data are available, ecological comparisons can be carried out [64a], [64b].

2.4.1. Technical Limitations

The technology for improving a process by avoiding or reducing residues is not always available. Each process must be researched and developed individually. Additional fundamental knowledge of chemical processes must be acquired, which requires time and money. Even so, the production of residues cannot be prevented completely. The idea of a production process with 100% yield, i.e., with "zero emission of hazardous materials of any kind [65]," is a vision that can never exist in reality [33], [66].

Also, the utilization of residues (e.g., contaminated solvents or solvent mixtures) may be limited for technical or economic reasons. Some practical examples of this are given below.

Volatile Organic Compounds in Poly (Ethylene Terephthalate) Synthesis [67]. In the production of poly(ethylene terephthalate), PETP, low boiling organics are produced during solvent processing. These consist of a mixture of methanol, methyl esters of aliphatic carboxylic acids, and cyclic ethers. No internal or external method of reusing this material could be found. A process is currently under investigation for hydrolyzing the methyl esters and recovering the methanol.

Waste-Gas Purification by Scrubbing with Glycol Ethers [35]. In the pharmaceutical industry, certain waste-gas streams are purified by scrubbing with glycol ethers. The solvent mixture produced on desorption contains ca. 16 components that are impossible to separate.

Complex Solvent Mixtures [68]. Problems preventing the separation of complex solvent mixtures are summarized below:

1) The azeotropic behavior of solvent mixtures (e.g., methyl acetate – acetone – methanol) can prevent distillative separation
2) The thermal instability of substances dissolved in a solvent mixture can also prevent distillative separation
3) Solvent residues are needed for dissolving partially polymerized organic residues to produce a material of suitable consistency for pumping into an incinerator

2.4.2. Economic Limitations

Investment decisions for new production plants or changes in production methods cannot be made without taking into account the existing market situation and the question of future demand for the product.

The quality and profitability of a product must be ensured under given market conditions. In principle, a new production process can be put into practice only if the costs, including the costs of residue disposal, are less than those of the existing process [69]–[71]. For any investment decision [72],

$$C_{INT} < C_{ST} + C_{EOP}$$

where C_{INT} are the total costs of the new integrated technology; C_{ST} the costs of the standard technology; and C_{EOP} the costs of retrofitting with the additive environmental technology.

If the existing production plant has not yet reached the end of its economic life, the total costs C_{INT} must also include the sunk-cost. This is the basic economic principle. In the case of investment this is modified by use of the following parameters: Net present value (NPV = discounted cash flow in ten years minus investment), internal rate of return (IRR) and pay back period (PBP). Generally should be: NPV > investment, IRR > cost of capital and PBP $\leq$ 5 years.

Economic considerations can also limit the utilization of residues. An example is given in the following: In the proposed manufacture of a pharmaceutical product (investment: 2.5×10^6 DM, ca. $\$ 1.6 \times 10^6$), a mixture of ethanol, triethylamine, and water is produced. Investigations aimed at producing the ethanol and triethylamine in a purity suitable for utilization showed that recovery of the ethanol and triethylamine would require a special distillation plant consisting of two specially designed columns and a special storage to comply with German regulations for flammable liquids. The investment of this plant would be 3.4×10^6 DM (ca. $\$ 2.25 \times 10^6$). The costs of the recovered solvents would be greater by a factor of 2.8 than the cost of new ethanol and triethylamine plus incineration of the residue mixture in the internal waste incineration plant. The solvent mixture was therefore disposed of by incineration [68].

2.5. Effect of Production-Integrated Environmental Protection

The effect of production-integrated environmental protection is represented by the amounts of residues prevented or utilized. These are compared with the amounts of residues not utilized — i.e., waste products. Figures have been provided by BASF, Bayer, and Hoechst. These are given below:

BASF. In 1995 810 100 t of products were manufactured from residues. The amount of material recycled within a process is considerable (e.g. recycling of raw material not reacted on passing through the reactor or recycling of auxiliares such as solvents). The quantities certainly exceed the quantities of products sold (1995: 8.1×10^6 t). By the introduction of new processes or the improvement of existing processes in the last 20 years, the formation of ca. 1.7×10^6 t of waste materials was prevented [73].

BASF operates a waste materials exchange system involving the internal or external supply or sale of by products, superseded products etc.

The quantity of residues utilized by outside companies in 1995 was 100 000 t. These included mainly the following:

1) Calcium carbonate from water treatment was used as a raw material in the cement and limestone industry.
2) Plastic waste was sold to external consumers.
3) Chlorinated hydrocarbons were used in the production of hydrochloric acid.
4) Ash from the sewage sludge incinerator was utilized in mining operations.

In 1995, the amount of waste produced was 201 000 t.

Bayer [74]. In the Bayer group (world-wide), ca. 750 000 t of production residues were utilized in 1996. Ca. 1.35×10^6 t of residues had to be disposed of as waste. Thus the fraction of the total residues that were utilized was 36%. The amount of residues produced in the same year was 2.1×10^6 t.

These figures for BASF and Bayer do not include internally recycled solvents. Therefore, they cannot be compared with the figures given for Hoechst below.

Hoechst. In the Hoechst group in Germany, ca. 2.3×10^6 t (2.8×10^6 t) of residues have been produced in manufacturing operations in 1995 (1994). By the introduction of new processes or the improvement of existing processes in earlier years, the formation of ca. 600 000 t of residues and waste materials was prevented [75]. 1.9×10^6 t (2.4×10^6 t) were recycled and utilized, and ca. 400 000 t (400 000 t) of residues had to be disposed of as waste. Thus, the fraction of the total residues that were utilized was 83% (85%) (in the solvent residue area: ca. 95% [67]), and 17% (15%) was disposed of as waste.

The most important residue groups are listed below, together with the recycled amounts in tonnes (in 1994):

Aliphatic and aromatic solvents by redistillation	883 000
Hydrochloric acid (31%) by distillative processing	313 000
Hydrochloric acid (100%) from vinyl chloride production used for oxychlorination	175 000
Chlorinated hydrocarbons by distillation	200 000
Energy-rich residues used as substitute fuels	58 000
Hydrochloric acid (31%) from incineration of chlorination residues	90 000
Sulfuric acid by distillation reprocessing or as supply to other users	60 000
Carbon disulfide from waste gas	13 000

Na$_2$SO$_4$ and (NH$_4$)$_2$SO$_4$ from wastewater and spinning baths	70 000
Recovery of acetic acid and acetaldehyde	85 000
Aromatic and aliphatic hydrocarbons	70 000
Activated carbon and carbon residues	67 000
Catalysts and inorganic substances	18 000
Utilization of plastic waste	55 000

2.6. costs of Integrated Measures

Capital investment in integrated environmental protection as a percentage of total investment for environmental protection is considered by economic experts to be an indicator of the quality of the environmental protection (EP) or an indicator of preventive environmental protection. The following statements have been made based on this indicator [76]–[79]: A "quantity–quality trade-off" occurs in the industry.

1) That means that a relative increase in the quantity of capital investment for end-of-pipe instead of production-integrated measures leads to a reduction in the overall quality of environmental protection
2) Capital investment into end-of-pipe technologies does not satisfy the precautionary measures preferred by environmental policy
3) Political support for end-of-pipe measures is short-sighted

All these conclusions are based on an interpretation of data from the Statistisches Bundesamt zur Grundstoff- und Produktionsgüterindustrie (German Federal Bureau of Statistics for the Basic Raw Materials and Capital Goods Industries). This interpretation is as follows [76]–[79]:

1) Integrated EP capital investment expressed as a percentage of total EP capital investment is ca. 20 %
2) Most capital investment is in end-of-the-pipe and not in integrated technology

However, the figure of 20 % has not been proved for the chemical industry. By using the author's data, the proportion of integrated EP investment in the chemical industry is calculated to be ca. 6 %, but the data cannot be verified in practice.

Several authors [80]–[82] agree with the above statements. However, these cannot be confirmed because annual statistics from the German Bureau of Statistics do not distinguish between additive and integrated measures [26]. Thus, sufficiently accurate information on such protective measures cannot be obtained from the statistics. This also applies to the proportion of integrated technologies in the total environmental protection capital investments for the Federal Republic of Germany, which was estimated to be 18 % in 1984 [27], [83].

Process costs following process redesign and utilization of residues can be calculated in detail using MFA and KFA. Technically these measures can be seen in terms of environmental impact as integrated environmental protection. At present there is no method that allows identification of the proportion of the investment and operating costs due to production-integrated environmental protection [54], [84]. The majority of these technical measures, however, are only carried out, when typical for every innovation, they reach economic targets:cost reduction and acceptable return on investment. The assessment also fails for new products made by new environmentally-sound processes.

Therefore, figures for integrated EP capital investment and resulting operating costs are not yet available. They are contained in the overall amounts of capital investment and operating costs of the production process. Moreover, the reduction in residues and emissions is a much better indicator of successful environmental protection than the amount of capital hypothetically invested [66].

3. Examples of Production-Integrated Environmental Protection in the Chemical Industry

3.1. Introduction

The following examples from the international chemical industry illustrate the concept of production-integrated environmental protection described in Chapter 2. Other publications and company monographs of recent years show that the total number of examples of this type in chemistry is actually much larger [31], [36], [37], [67], [68], [85]–[88].

However, the uneven distribution of these examples with respect to countries shows that so far not all nations place equal emphasis on production-integrated environmental protection in the chemical industry. This observation may be due to the following factors:

1) Differences in legal requirements limiting emissions and residues from production processes
2) The intensity of research required for production-integrated process improvements
3) The close association between production-integrated measures and investment decisions on new production plants or essential modification of processes
4) The usually higher degree of complexity of production-integrated methods compared to conventional processes with additive environmental protection
5) The dependence on the quality of raw materials available when certain components or impurities associated with the raw materials are decisive for emission and residue problems
6) The influence of the operation mode: continuous processes being simpler to optimize than batch processes

The examples given here can be classified in accordance with one or more of the following principles:

1) Prevention or reduction of emissions and residues by modification or optimization of chemical syntheses:
   – Development and use of novel and more selective catalysts. Examples: 3.2.1.10., 3.2.1.11., 3.2.2.1, 3.2.2.3, 3.2.7, 3.2.10
   – "Biocatalysis" Examples: 3.2.1.9, 3.2.14.1, 3.2.14.2
   – Kinetic optimization of process conditions. Examples: 3.2.5.2, 3.2.7, 3.2.8
   – Increase of product yield by shifting of the equilibrium and product recycling. Examples: 3.2.1.7, 3.2.2.2, 3.2.5.1, 3.2.7

- Nonaqueous synthesis route. Examples: 3.2.1.3, 3.2.2.1 (intermediate product), 3.2.10
- Low-residue digestion of raw material. Example: 3.2.6
- Prevention of by-products by avoiding associated substances in the raw materials. Examples: 3.2.5.2
- Change to continuous operation. Examples: 3.2.1.8, 3.2.2.1, 3.2.7

2) Prevention of emissions by optimized operation techniques

- Low-emission vacuum technique. Example: 3.2.1.5
- New methods of product separation and purification. Examples: 3.2.2.2, 3.2.2.5, 3.2.13

3) Process-integrated recycling of auxiliaries and by-products. Examples: 3.2.1.1, 3.2.1.2, 3.2.1.8, 3.2.1.12, 3.2.2.4
4) Recovery of raw materials or auxiliaries for the production process by reprocessing of residues. Examples: 3.2.1.4, 3.2.1.6, 3.2.2.4, 3.2.12, 3.2.13
5) Utilization of residues and joint products by linking of production processes. Examples: 3.2.1.1, 3.2.1.4, 3.2.1.6, 3.2.4, 3.2.11
6) Energy saving and prevention of emissions due to burning of fossil fuels by utilization of process energy, recovery of energy from residues, and higher process efficiency. Examples: 3.2.1.3, 3.2.1.8, 3.2.1.9, 3.2.2.6, 3.2.3.1, 3.2.5.2, 3.2.7, 3.2.8, 3.2.9, 3.2.10, 3.2.11, 3.2.14.1, 3.2.14.2

3.2. Selected Examples

3.2.1. Examples from Hoechst

3.2.1.1. Recovery and Utilization of Residues in the Production of Viscose Staple Fiber

General Features of Viscose Staple Fiber Production [89]. Cellulose from wood is the most important starting material for staple fiber production by the viscose process. Cellulose xanthate is obtained by converting cellulose to sodium cellulose followed by treatment with carbon disulfide.

Reaction scheme
1) Production of alkali cellulose
   Cellulose + NaOH → Alkali cellulose
2) Aging

3) Viscose formation (dissolution of alkali cellulose by addition of carbon disulfide)

$$\text{Alkali cellulose} + CS_2 \rightarrow \text{Cell-O-C}\begin{array}{c}=S\\|\\SNa\end{array}$$

Side reaction:
$$2\,CS_2 + 6\,NaOH \rightarrow Na_2CS_3 + Na_2CO_3 + 3\,H_2O + Na_2S$$

4) Ripening
5) Regeneration of cellulose fibers in the $H_2SO_4 - Na_2SO_4 - ZnSO_4$ coagulation bath
$$\text{Cell}-O-CS_2Na + NaHSO_4 \rightarrow Na_2SO_4 + CS_2 + \text{Regenerated cellulose}$$
Side reactions:
$$Na_2CS_3 + H_2SO_4 \rightarrow Na_2SO_4 + H_2S + CS_2$$
$$Na_2S + H_2SO_4 \rightarrow Na_2SO_4 + H_2S$$
$$Na_2CO_3 + H_2SO_4 \rightarrow Na_2SO_4 + H_2O + CO_2$$

Cellulose xanthate is dissolved in aqueous sodium hydroxide solution to give a viscous liquid—the viscose. The sodium hydroxide solution, which is added in excess, also results in the formation of by-products such as sodium trithiocarbonate and sodium sulfide. In addition sodium carbonate is formed.

The ripening of viscose is stopped after a certain time, depending on the nature of the required end product, and the cellulose is precipitated. This is performed by injecting the filtered spinning solution through fine nozzles (the orifices of the spinneret) into an acid bath. The main reaction occurring in the acid bath is the coagulation of viscose, and CS_2 is formed. In side reactions, salts formed in the preparation of the spinning bath decompose to H_2S, CO_2, and Na_2SO_4 (step 5).

As the strand or slubbing is led out of the spinning bath, subsequent degassing occurs, in which carbon disulfide and hydrogen sulfide dissolved in the acid spinning bath partly evaporate. The spun fibers then pass through several aftertreatment baths. In the course of this, adhering sulfuric acid is washed off. A subsequent degassing of carbon disulfide and hydrogen sulfide also occurs in the wash baths, which are mostly hot.

For production of 1 t of staple fiber, up to 350 kg of CS_2 is required. About 25 % of the CS_2 decomposes to H_2S and CO_2. The schematic of the entire process is shown in Figure 14.

The Problem. During the production of staple fibers by the viscose process, toxic *waste gases* containing carbon disulfide and hydrogen sulfide are formed. In the conventional process, those waste-gas streams that contain mainly carbon disulfide were collected in a waste-gas system with associated recovery. These waste gases arise during deaeration and exhaustion in viscose production and during aftertreatment of the viscose staple tow. In the recovery plant, after the separation of the hydrogen sulfide fraction, carbon disulfide was adsorbed on activated carbon. Up to about 65 % of the carbon disulfide used was recovered.

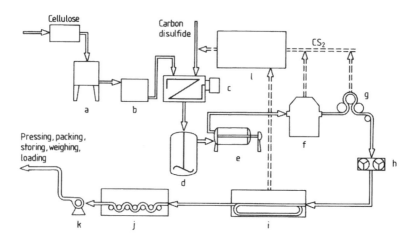

Figure 14. Schematic of viscose staple fiber production
a) Mashing; b) Ripening; c) Sulfidization; d) Dissolving; e) Filtration; f) Spinning; g) Stretching; h) Cutting; i) Washing and aftertreatment; j) Drying; k) Transport; l) CS_2 recovery

The greater part of the other sulfur-containing waste gases — ca. 60–90 kg of sulfur in the form of CS_2 and H_2S per tonne of staple fiber — previously entered the waste air. This part originated mainly from the spinning process and also from the spinning bath components discharged with the fibers, from which the gases are released by diffusion in the later treatment stages. In the original procedure, the waste-gas sources in the process had to be exhausted with very large quantities of air to keep waste-gas concentrations safely below the in-plant threshold limit. The specific quantity of waste gas was therefore extraordinarily high (ca. 235 000 m^3/h at an output of 55 000 t of viscose staple fiber per year). The extracted gases, greatly diluted, reached the open air via a tall stack. As a result, nuisance odors often occurred in the surroundings of viscose staple fiber plants because the threshold odor concentration of hydrogen sulfide, at ca. 0.005 mg/m^3, is extraordinarily low. Disposal of the waste gas failed because of its large volume and low concentration.

Considerable quantities of *sodium sulfate* and *zinc sulfate* are also formed during regeneration of cellulose fibers. Part of the sodium and zinc sulfates is discharged from the spinning bath together with the viscose fibers and enters the wastewater through the aftertreatment of the fibers.

Solution [89]–[92]. To solve the waste-gas problem, a recycling scheme was developed by Hoechst and Süd-Chemie, Kelheim/Germany, and realized in 1975–1979. In this process, 80% of the viscose waste gases were collected. In the following years, further increase of collection to the current value of 90% was achieved, and the waste-gas volume was simultaneously reduced considerably. In the recycling scheme, sulfur-containing waste gases from viscose staple fiber production (Hoechst) are supplied to a neighboring sulfuric acid plant (Süd-Chemie) where they are used as a raw material or as combustion air. The sulfur compounds contained in the waste gases are combusted to SO_2, CO_2, and H_2O, and the SO_2 is processed to H_2SO_4. The sulfuric acid produced can be used again directly for the coagulation bath. The new process concept required substantial changes and new developments in both plants. In addition it provided the preconditions for improved recovery of carbon disulfide in the viscose plant.

New Concept of Waste-Gas Capture during Viscose Staple Fiber Production. As a result of the basic chemical reactions in a sulfuric acid plant and the desired conversion of nearly 100%, the ratio of the quantity of combustion air to sulfur is fixed. If the entire amount of waste gas formed in a conventional viscose staple fiber plant were to be used as combustion air in a sulfuric acid plant, about 30 times as much sulfuric acid would have to be produced as the staple fiber plant consumes. The quantity of lean waste gas from the viscose plant must therefore first be matched to the much lower combustion air requirement of the sulfuric acid plant. This can be achieved by appropriate reduction of the air exhaustion in the viscose staple fiber plant. However, this is possible only if, at the same time, risks to safety as a result of exceeding the in-plant threshold limit value and of the occurrence of explosive gas–air mixtures are avoided.

For this purpose, waste gases from the various process stages are assembled in two separate systems: waste-gas system 1 ("rich gas") is used for the collection of combustible but not explosive gases (waste-gas concentration >explosive range), and waste-gas system 2 ("lean gas") is used for the collection of waste gases of low concentration (waste-gas concentration <explosive range). Extensive changes to the viscose staple fiber plant were necessary to minimize waste-gas emission:

1) The quantity of carbon disulfide required in viscose production was reduced by developing a special aging process.
2) Because of the new concept of collection of waste air from viscose production, the spinning machines, stretching and tow washing, and subsequent waste-air treatment, today most of the carbon disulfide can be recovered from the waste gas.
3) To recover hydrogen sulfide from the spent spinning bath acids as completely and at as high a concentration as possible, a vacuum degassing plant was constructed. The concentrated waste-gas stream of ca. 200 m^3/h can be utilized directly in the sulfuric acid plant (rich gas).
4) By careful enclosure of the spinning machines and the apparatus in the acid station and by using novel exhaustion systems, the total waste-air stream of 180 000 m^3/h (STP) has been reduced to 10 000 m^3/h (STP). The entire spinning machine waste air can therefore be routed via the waste-gas purification plant (CS_2, H_2S). Lean gas from the acid station is used as combustion air in the sulfuric acid plant.
5) Finally, an extensive waste-air collection and piping system had to be constructed. Corrosion problems in the lines had to be ruled out by selecting suitable materials.
6) To ensure safety, a series of additional measures was required: e.g., a largely flangeless construction and reduced-pressure mode of operation of the waste-gas system to avoid leaks; installation of a process measuring and control technology with emphasis on safety; and construction of an additional exhaustion system on the spinning machines for manual operating procedures.

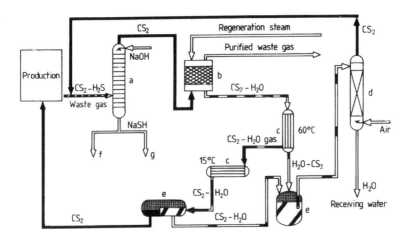

Figure 15. Carbon disulfide recovery in viscose staple fiber production
a) Wash column; b) Adsorption on activated carbon; c) Condenser; d) CS_2 stripper; e) Separator; f) Zinc precipitation; g) Acidification (H_2S used for H_2SO_4 production)

Hydrogen Sulfide Recovery. Waste air from the production installations (spinning, stretching, tow washing), which is contaminated with CS_2 and H_2S, is freed from H_2S by a multistage wash with sodium hydroxide solution (Fig. 15, a). In the course of this, an NaSH-containing spent lye is formed. Subsequently, CS_2 contained in the waste gas is adsorbed on activated carbon (b) and then desorbed with steam (Supersorbon process). The desorbate water –CS_2 vapor is precondensed at 60 °C (c). The gaseous residue is then condensed at 15 °C, and the condensate is drained off into the separator (e). The aqueous phase obtained, together with the precondensate, is stripped of CS_2 in a column (d). The CS_2 is returned to the production process.

Zinc Recovery. Part of the spent lye from the gas wash (a) is used to precipitate zinc (f) from the stretching bath and from the wastewater of the tow wash. By addition of H_2SO_4 to the precipitated zinc sulfide, zinc sulfate is recovered for the spinning process. During zinc precipitation, a small portion of zinc polysulfide is formed that cannot be converted to zinc sulfate. This portion must be discharged.

Hydrogen Sulfide Utilization. The H_2S content of the remaining spent lye (g) is released by adding H_2SO_4 and utilized in the neighboring sulfuric acid plant together with H_2S from zinc sulfate recovery and waste gas (CS_2, H_2S) from the spinning bath regeneration.

Sodium Sulfate Recovery. In the spinning bath regeneration, the spinning bath is partially evaporated. In the course of this, sodium sulfate is obtained in a marketable form by crystallization. The overall flow sheet of the new process is shown in Figure 16 A.

Supplementary Engineering Equipment for Process Rearrangement. To be able to utilize the lean gas stream, which is saturated with water vapor, in the sulfuric acid plant, a new wet-dry catalytic process for sulfur combustion was developed jointly by

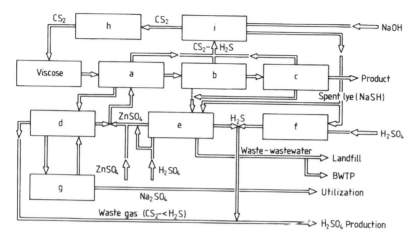

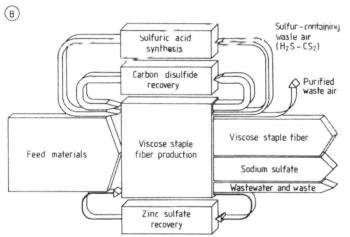

Figure 16. Residue utilization in viscose staple fiber production
A) a) Spinning; b) Stretching; c) Tow wash; d) Spinning bath regeneration; e) Zn recovery; f) Spent lye degassing; g) Na_2SO_4 crystallization; h) Recovery; i) WashingBWTP = Biological wastewater treatment plant
B) Overall balance of the integrated pollution control solution

Süd-Chemie and Lurgi. In this process the water vapor from the $SO_3 - SO_2$ gas mixture is removed after the third catalytic stage by condensation in a new absorption process. The dry $SO_3 - SO_2$ gas mixture then passes through the fourth catalytic stage. Depending on the degree of dilution of the waste gases (with N_2 and CO_2 as inert constituents), the new process gives a sulfuric acid yield of 99.1 – 99.4 %.

Pollution Control Balance. The overall balance of the integrated pollution control is shown in Figure 16 B. As a result of collecting the waste-air streams, ca. 90 % of the original emissions of sulfur compounds from the plant can be avoided, corresponding to ca. 5000 t/a (calculated as S). Only the waste air arising during manual operations on the spinning machines, which is greatly diluted as a result of the additional exhaustion, must be discharged from the stack as before. Apart from rare plant upsets, the nuisance odor in the surroundings of the plant is a thing of the past.

Moreover, during spinning bath regeneration, about 40 000 t/a of anhydrous sodium sulfate is recovered. This corresponds to about one-half of the amount formed in the overall viscose process (including neutralization steps). The remaining wastewater enters a Biohochreactor for degradation of the organic materials (dissolved cellulose ingredients).

The solution found here to the viscose waste-gas problem dispenses with additional energy requirement in the form of electric power, fuel gas, or natural gas. No new waste materials are produced. The operating costs of the process depend basically on the market price of sulfuric acid. The process scheme has by now been applied internationally in several viscose plants.

3.2.1.2. Recovery of Methanol and Acetic Acid in Poly (Vinyl Alcohol) Production [35], [93]

In the production of poly(vinyl alcohol), a mother liquor containing methanol and methyl acetate is produced. This mother liquor is converted to acetic acid and methanol in a multistage process; these substances are then recycled to the production process.

In the first stage, the mother liquor is subjected to distillation. An azeotropic mixture of methanol and methyl acetate is obtained as overhead product. The methanol–water mixture, which remains in the distillation bottom, is fed to a second distillation stage, and the methanol distilled off is recycled to the production process.

The azeotropic mixture is subjected to an extractive distillation (extraction medium: wastewater from the methanol distillation). The still bottoms so obtained, a water–methanol mixture, are fed to the methanol distillation. Methyl acetate taken from the top of the column is fed to a reactor where it is hydrolyzed to methanol and acetic acid with the aid of an ion-exchange resin. In the subsequent distillation, acetic acid is removed, and the mixture of methanol and unreacted methyl acetate from the extractive distillation is recycled (Fig. 17). Wastewater from this recovery process is treated in a biological wastewater treatment plant.

3.2.1.3. Acetylation without Contamination of Wastewater [94]

The acetylation of 2-naphthylamine-8-sulfonic acid was formerly carried out in an aqueous medium with acetic anhydride. The product, 2-acetaminonaphthalene-8-sul-

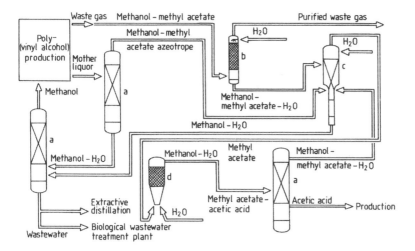

Figure 17. Recovery of methanol and acetic acid
a) Distillation; b) Scrubber; c) Extractive distillation; d) Hydrolysis reactor

fonic acid, was salted out from the aqueous medium by addition of ammonium sulfate and filtered off.

The mother liquor together with the washwater from this filtration contained 1200 kg of ammonium sulfate (COD 1000 kg) for each 1000 kg of product. This wastewater was treated in a biological wastewater treatment plant (see Figs. 18 and 19). The formation of this highly contaminated wastewater was the reason for replacement of the process by dry acetylation. For this, a flat-bottomed vessel with stirring equipment was installed (Fig. 20, a). Because of the dry process, no contaminated wastewater was formed; thus, treatment of ca. 2000 kg of sewage sludge in the wastewater treatment plant was avoided. In addition, 270 kg of acetic acid was recovered per 1000 kg of product.

3.2.1.4. Reutilization Plant for Organohalogen Compounds

Introduction. Organic halogen compounds, especially those containing bromine and fluorine, are intermediates in the synthesis of pharmaceuticals, plant-protection agents, special high-performance plastics, and liquid crystals for optical displays. These intermediates must be produced in highly pure state. They are therefore separated from by-products formed during the chemical synthesis in later purification stages.

Since the by-products are often not directly reutilizable, they are disposed of in a waste incineration plant. Up to now only a recovery of the energy content is possible; a chemical reutilization of the by-products is not. Purification of the flue gas does not lead to reutilizable sustances.

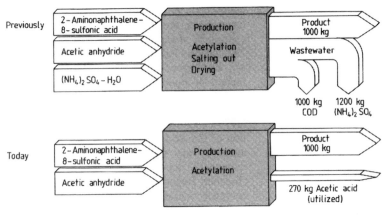

Figure 18. Comparison of previous and current production processes of 2-acetaminonaphthalene-8-sulfonic acid

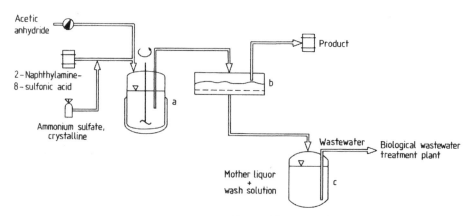

Figure 19. Production of 2-acetaminonaphthalene-8-sulfonic acid by wet acetylation
a) Acetylation and salting out; b) Filtration; c) Separation

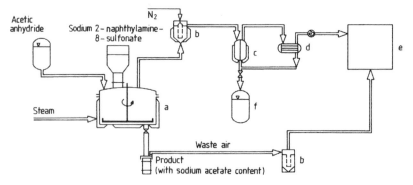

Figure 20. Production of 2-acetaminonaphthalene-8-sulfonic acid by dry acetylation
a) Acetylation and drying; b) Filtration; c) Cooling trap; d) Condenser; e) Central waste-air purification plant; f) Separation

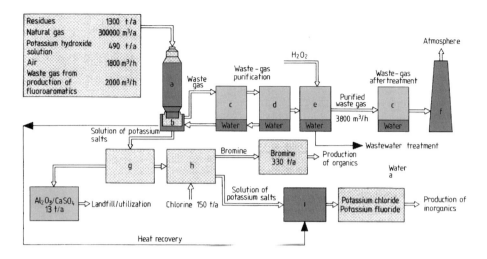

Figure 21 Recovery of bromine from residues containing brominated organics
a) Incineration and salt production (1200 °C); b) Quench cooling (85 °C); c) Scrubber (two stages); d) Electrostatic filter; e) Scrubber (one stage); f) Stack; g) Filtration; h) Bromine recovery; i) Evaporation

The objective was not only to utilize the calorific values of these by-products or residues but also to reuse them chemically. This was achieved by thermal oxidation (incineration) with simultaneous neutralization to give potassium halides that are subsequently dissolved in water. The salt solution obtained is used for the recovery of potassium bromide, potassium fluoride, and potassium chloride as raw materials.

Method of Operation (Fig. 21). Liquid organic residues, which contain mainly bromine compounds with some chlorine and fluorine compounds, are fed through jets into a vertical cylindrical oxidation chamber (incinerator, a). The essential feature of the oxidation chamber is a special burner system that causes intensive turbulence of liquid and gaseous residues, fuel, and combustion air. An auxiliary flame produced by a special burner fired with natural gas is necessary to establish and ensure continuous operation of the oxidation process. This burner produces a very intensely radiating flame that reduces the mixing zone within the oxidation chamber. Intensive mixing of the gases ensures complete reaction, which results in comparatively low emissions of NO_x and CO. The high calorific value of the organohalogen residues ensures that the thermal energy necessary for the process is liberated.

Waste air from the production of halogenated aromatics is used to form part of the combustion air, so that the residues present in the waste air (e.g., fluorine compounds) can also be treated in the recovery process.

The residues are converted to hydrogen halides in the oxidation chamber at 1200 °C and a residence time of at least 2 s. The hydrogen halides are then immediately neutralized by the simultaneous spraying of potassium hydroxide solution into the chamber, to form potassium halides. This prevents the formation of free hydrogen halides or of elementary halogens according to the Deacon equation. In the case of bromine, the equation

$$2\,Br_2 + 2\,H_2O \rightleftharpoons 4\,HBr + O_2$$

is shifted completely to the right by immediate neutralization of the hydrobromic acid. The equilibrium reaction is then no longer reversible, and complete conversion of the residues to hydrogen halides and, ultimately, potassium halides occurs.

This also avoids the secondary reaction

$$Br_2 + 2\,KOH \rightleftharpoons KBr + KBrO + H_2O$$

which occurs if the potassium hydroxide solution is added to the waste gas at a later stage, as in conventional exhaust-gas purification.

The potassium salts that are liquid at >800 °C flow partially down the walls to the bottom of the oxidation chamber and are dissolved in the water bath in the quenching vessel (b) directly below. The vapor-phase fraction of the salt forms aerosols in the quenching vessel on cooling to ca. 85 °C. The concentrations of these aerosols are then reduced in the multistage waste-gas purification equipment (c–e) to levels below the limit values stipulated by Regulation 17 of the German Federal Antipollution Law (17. Verordnung Bundesimmissionsschutzgesetz 17. BImSchG). In this equipment, waste gas is passed successively through two jet scrubbers (c), a condensation electrostatic filter (d), and a packed column fed with H_2O_2 (e).

The liquid level in the quenching vessel is kept constant by adding liquor from the waste-gas scrubbers. Thereby some evaporation of the salt solution occurs.

Insoluble components of the salt liquor (ca. 1% of the residues treated) are removed by filtration. These components consist mainly of insoluble inorganic salts and metal oxides; they can be dumped or, preferably, reutilized.

Elementary bromine is obtained from the filtered salt solution by reaction with chlorine in a further process step. Potassium chloride and potassium fluoride are recovered by evaporation of the remaining solution (with heat recovery from the quenching vessel) and are internally recycled.

The purified waste gas (see above) then undergoes a final purification by passing through a two-stage waste-gas scrubber in a production plant for inorganics (see Fig. 21).

Advantages of the Process. The formation of dioxins is prevented by the high oxidation temperature (1200 °C), the residence time of 2 s in the reactor, the mainly homogeneous temperature profile in the oxidation chamber, and the quench cooling of the reaction gases from 1200 to 85 °C. Measured concentrations are <0.1 ng/m^3 toxicity equivalents dioxins and furans.

Known alternative methods for treating liquid residues either have disadvantages or offer no advantages:

1) *Hydrogenation process* The high fluorine content of the residues leads to problems with construction materials. Elementary bromine is not recovered. Production of the hydrogen required for the process is energy intensive.
2) *Plasma process* Since the plasma burner can be used only as a pre-stage before the subsequent oxidation with air or pure oxygen, the consumption of electrical energy, and the cost of the plant are excessively high. Further recovery of the halogens will be similar to that used in the recommended process. The ecological balance of this process is therefore very unfavorable.
3) *Pressure gasification* A mixture of CO, CO_2, and hydrogen halides is formed at ca. 1600 °C. The hydrogen halides must be separated by waste-gas purification to recover the halogens. The formation of elementary bromine in the gasification process is problematic and would lead to considerable difficulties in waste-gas purification.
4) *Chemical reverser* In a converter tube clad with a special ceramic, a reductive atmosphere with a hydrogen excess at very high temperature is produced by combustion of propane under pressure. The fluid organohalogen compounds are decomposed in this temperature region.

For the substances mentioned above containing high levels of aromatics, the mixture would have to be led through a combustion chamber before entering the reactor to convert the molecules into suitable fragments that are able to react with hydrogen in the cracking process.

The later process steps, such as quenching and gas purification, would be very similar to those in the process initially described. It would be complicated to include the use of waste air as combustion air or the use of heat of reaction in the concept of the chemical reverser.

3.2.1.5. Vacuum Technology for Closed Production Cycles

In a chemical production plant, a large number of different machines, fittings, and safety equipment must work together reliably. With suitable engineering design, the amount of residues can be reduced or the residues can be recovered.

In the chemical industry, many processes are operated at reduced pressure (between 2000 Pa and atmospheric pressure). Liquid ring pumps, which are displacement pumps, are widely used for pressure reduction. The working liquid of a liquid ring pump must perform the functions of a piston, a seal, and a heat absorber.

This working liquid can be water or an organic liquid. Water ring pumps are used in recycling operations to reduce wastewater contamination. If a gas at higher temperature saturated with a solvent is pumped, the same solvent can be used as the working liquid of the pump. In a recycling operation, such a system can be designed for solvent recovery by connecting it to a rectification column. An example of this system is shown in Figure 22, in which isopropanol is the solvent.

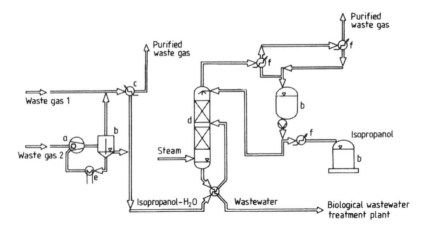

Figure 22. Vacuum plant with liquid ring pump
a) Liquid ring pump; b) Separator; c) Waste-gas condenser; d) Rectification; e) Cooler for working liquid (isopropanol); f) Condenser

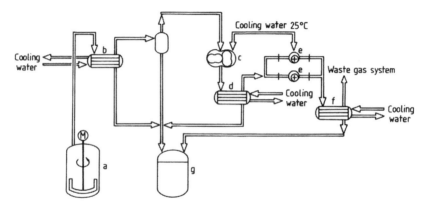

Figure 23. Vacuum plant with nonlubricated sliding vane vacuum pumps to reduce pollution of wastewater
a) Reactor; b) Precondenser; c) Roots vacuum pump; d) Intermediate condenser; e) Sliding vane rotary pump (two stage); f) Final condenser; g) Condensate collection vessel

Such a system cannot be used for pumping solvent mixtures or low-boiling solvents (e.g., acetone). Nor can water ring pumps be used because these lead to contamination of the wastewater. Here, sliding vane rotary vacuum pumps (without oil lubrication) can be used. Such a vacuum pump has been developed for pumping solvent mixtures produced during synthetic resin production. It has been put into operation by Herberts in association with the company Provac (see Fig. 23). The following plant design was used: The vapors are drawn under reduced pressure through the precondenser (b) to the Roots vacuum pump (c). The gases that have been heated by compression in the Roots vacuum pump are then cooled in the intermediate condenser (d), and the condensate formed is automatically removed via a condensate trap into a collection vessel (g). The gases leaving the intermediate condenser are fed into two sliding vane rotary vacuum pumps (e) arranged in parallel. The gases from these pumps are fed into a condenser (f) in which more condensate is formed; this also drains into the collection vessel. The residual vapors are then fed to the waste-gas purification system. This method of operation reduces contamination of the wastewater, and a mixture of solvents is recovered. Furthermore, the inclusion of a Roots vacuum pump eliminates the use of a steam ejector pump, thereby avoiding the contamination of wastewater that such a pump would cause.

3.2.1.6. Utilization of Exhaust Gases and Liquid Residues of Chlorination Processes for Production of Clean Hydrochloric Acid

When organic compounds are chlorinated on an industrial scale, large quantities of hydrogen chloride are formed. This is contaminated with chlorinated hydrocarbons formed as by-products of the chlorination process.

A measure of the degree of contamination is given by the content of halogen compounds that can be adsorbed from the acid by activated carbon (AOX content). In nearly all chlorination reactions the chlorine is incompletely consumed, so that a fluctuating proportion of free chlorine usually occurs both in the hydrochloric acid recovered from the residual gases from the chlorination process and in the waste gases.

This free chlorine is as undesirable as other impurities in both the waste gas and the hydrochloric acid, and must be removed for environmental reasons and to improve acid quality.

One method of producing cleaner hydrochloric acid and a waste gas with only traces of organic impurities is to subject the chlorinated residues to thermal treatment before the absorption stage. In this treatment, the impure, hydrogen-chloride-containing gases are heated in suitable combustion chambers to ca. 600–1200 °C. The energy is usually provided by the combustion of natural gas or other fuels in an excess of oxygen, the temperature being adjusted to suit the impurities present. The flue gases obtained contain low levels of organic substances as well as some free chlorine. The latter occurs because in the presence of oxygen, free chlorine originally present in the chlorination residues does not react quantitatively to form HCl according to the equation

$$2\,Cl_2 + 2\,H_2O \rightarrow 4\,HCl + O_2$$

Instead, additional free chlorine can even be formed.

In subsequent absorption of the hydrogen chloride, free chlorine partly dissolves in the hydrochloric acid formed and partly remains in the waste gas. Removal of chlorine from both the hydrochloric acid and the waste gas is technically difficult. Air blowing or stripping the hydrochloric acid gives incomplete removal of the chlorine and causes a considerable decrease in the concentration of the acid.

Moreover, additional chlorine-containing and hydrogen-chloride-containing waste-gas streams are formed, and these too must be purified. Like the dechlorination of the original waste gases, this can in principle be performed only by alkaline scrubbing, which transforms the waste-gas problem into a wastewater problem.

A method for removing hydrogen chloride and chlorine from combustion gases with recovery of hydrochloric acid in a multistage gas scrubbing process has been described [95].

In another process for reducing the free chlorine content of flue gases from a combustion plant [96], hydrogen is fed into the quench zone of the combustion apparatus at a rate that depends on the air rate to the combustion chamber. By this means, the chlorine content of a combustion gas has been reduced from 600 to 48 mg/m^3. However, chlorine-free waste gases have not yet been produced from flue gases with such high chlorine contents.

Finally, a process has been developed for disintegrating chlorinated hydrocarbon residues with recovery of hydrochloric acid [97]. The chlorine content in the flue gas from the combustion process is kept as low as possible by minimizing the oxygen content and feeding steam or water into the combustion chamber. However, the hydrochloric acid produced is apparently not completely free of elementary chlorine, because chlorine is sometimes still found in the waste gas after HCl absorption.

Methods are also described for the combustion of liquid and solid residues with the production of hydrochloric acid [98], [99]. In these processes also, the starting material is burned in an excess of oxygen, as already described. Thus, these methods have the same disadvantages as those mentioned, because chlorine is present both in the hydrochloric acid produced and in the waste gas.

A method has been developed for producing hydrochloric acid with low levels of AOX and almost no elementary chlorine, together with a waste gas that contains only traces of these substances and can be released directly into the atmosphere. Here, the hydrogen chloride and chlorine-containing crude gases contaminated with organic compounds are thermally treated by using hydrogen or hydrogen-containing combustion gases with a deficiency of oxygen. The fuel and air rates are adjusted so that no carbon is formed in the combustion zone and an excess of hydrogen is present in the exhaust gas [100].

Although this process enables chlorination residues of the usual composition to be purified so that they can be used to produce pure hydrochloric acid, in extreme cases (e.g., if very high levels of impurities are present in the crude gases or if liquids are to be treated), problems with carbon formation in the combustion chamber can arise. This is undesirable, because both the plant and the products would be fouled.

Novel Process [101], [102] (Fig. 24). A thermal process was needed in which the reaction would proceed in a stable and controllable fashion. This process would also have to be suitable for purifying both crude gases with extremely high levels of impurities and chlorine-containing liquids. These could be converted to hydrogen chloride so that the latter could be directly absorbed without further purification processes, to give chlorine-free hydrochloric acid and a waste gas suitable for immediate discharge to the atmosphere.

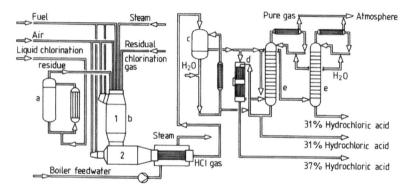

Figure 24. Production of pure hydrochloric acid from residual chlorination gases and liquid chlorination residues
a) Evaporator; b) Two-stage combustion; c) Quench; d) Isothermic absorber; e) Adiabatic absorber

The problem was solved by using thermal treatment of the impurities in a combustion operation with an excess of oxygen, and by following the first combustion stage by a second in which reaction gases from the first stage are subjected to further thermal treatment, in this case in a reducing atmosphere.

The crude gas mixture to be treated contains, in addition to hydrogen chloride and chlorine, considerable quantities of chlorinated hydrocarbons and other organic chlorine compounds. The mixture is subjected to a *two-stage thermal treatment*. It is fed into a *first combustion chamber* supplied with sufficient fuel and air. (Under some circumstances, replacing the combustion air either completely or partially with oxygen can be advantageous in this stage of the combustion process.)

Gaseous fuel (e.g., natural gas or hydrogen) is generally used in the first combustion stage, although liquefied petroleum gas or light or heavy fuel oil can also be used. In the *second combustion stage*, fuels with the lowest possible carbon content are preferred, because carbon formation is more likely here than in the first combustion stage since the process requires a deficit of oxygen.

If high-carbon fuels are used in the first combustion chamber, or if the raw gas contains large quantities of compounds with a high carbon content, the substoichiometric reaction in the second combustion chamber can lead to the formation of large amounts of carbon monoxide according to

$$CO_2 + H_2 \rightarrow CO + H_2O$$

After HCl absorption, the waste gas cannot be discharged directly because of legal restrictions. In this case, CO in the waste gas is converted to carbon dioxide, preferably in an afterburner under controlled conditions or sometimes catalytically.

The combustion chambers generally consist of steel vessels lined with a refractory ceramic material. Since this lining is not usually gastight, the highly corrosive raw gases, which contain hydrogen chloride and especially chlorine, can corrode the interior walls of the steel vessels at the prevailing surface temperatures, so that metallic ions, especially iron ions, appear in the hydrochloric acid formed. If high-purity hydrochloric acid is required (i.e., a product that not only is free of elementary chlorine and organic compounds but also contains only traces of inorganic compounds such as metallic salts, especially iron chlorides), the interior walls of the steel vessels and combustion chambers should be given a corrosion-resistant coating (e.g., enamel) before application of the porous ceramic material.

The crude gas to be treated is fed into the first combustion chamber. If necessary, liquids can first be passed through a vaporizer, and the vapors produced can then be mixed into the rest of the crude gases. If the liquids have high boiling points, feeding steam into the vaporizer can be useful. Combustion can be improved by adding combustion air, fuel, and additional steam or chlorine gas to increase the HCl content of the reaction gases.

The combustion temperature in the first chamber can be 800–1600 °C (preferably 1000–1300 °C); this is maintained by an auxiliary burner. The fuel–air mixture is adjusted so that an oxygen content of 0.1–11 vol % (preferably 0.5–5 vol %) is measured in the reaction gases at the exit of the first combustion chamber.

The reaction gases pass from the first to the second combustion chamber. There the temperature is maintained at 800–1600 °C (preferably 1000–1300 °C) by a burner supplied with a fuel –air mixture whose ratio is adjusted so that the reaction gases at the exit of the combustion chamber have a hydrogen content of 1–15 vol % (preferably 2–5 vol %).

Sometimes, the temperature of the second combustion chamber may have to be decreased by feeding steam into this chamber.

The reaction gases are passed through one or more absorption columns after cooling.

Preferably the pressure in the first combustion chamber should be kept constant automatically by a blower installed downstream.

The hydrochloric acid produced is practically free of elementary chlorine and organic chlorine compounds, and requires no further treatment. Also, the purified gas can be discharged directly into the atmosphere.

3.2.1.7. Production of Neopentyl Glycol: Higher Yield by Internal Recycling [75], [103], [104]

In the process for producing neopentyl glycol (2,2-dimethyl-1,3-propanediol) a joint product is formed and some starting materials remain unreacted. The latter are internally recycled to the process with the aid of the coupled products.

The first reaction, in reactor 1, is an aldol condensation in which formaldehyde is reacted with isobutyraldehyde (2-methylpropanal) in an alkaline medium with tri-*n*-propylamine as the catalyst to form hydroxypivalaldehyde (3-hydroxy-2,2-dimethylpropanal).

$$HC\overset{O}{\underset{H}{\diagdown}} + (CH_3)_2-CH-C\overset{O}{\underset{H}{\diagdown}} \rightleftharpoons CH_2(OH)-C(CH_3)_2-C\overset{O}{\underset{H}{\diagdown}}$$

The reaction mixture is transferred to a separating vessel where two layers are formed. The organic phase contains hydroxypivalaldehyde and an equilibrium amount of isobutyraldehyde. The heavier aqueous phase contains hydroxypivalaldehyde together with small amounts of unreacted starting materials (formaldehyde and isobutyraldehyde). The organic phase is hydro-genated in reactor 2 using a nickel contact catalyst. Here, the target product neopentyl glycol is formed from the hydroxypivalaldehyde and, isobutanol is formed from the isobutyraldehyde as the joint product.

$$H_2C-\underset{\underset{OH}{|}}{\overset{\overset{CH_3}{|}}{C}}-C\overset{O}{\underset{H}{\diagdown}} + H_3C-CH-C\overset{O}{\underset{H}{\diagdown}} \xrightarrow{H_2} H_2C-\underset{\underset{OH}{|}}{\overset{\overset{CH_3}{|}}{C}}-CH_2$$
$$\text{(Target product)}$$

$$+ H_3C-\underset{\underset{CH_3}{|}}{CH}-\underset{\underset{OH}{|}}{CH_2}$$
$$\text{(Joint product)}$$

Neopentyl glycol and isobutanol are separated by distillation. Some of the recovered isobutanol is used to extract the aqueous phase formed in the aldol condensation, and the remainder is used to produce synthesis gas. The isobutanol extract contains hydroxypivalaldehyde and starting materials. It is therefore recycled to the aldol condensation reactor. The isobutanol transferred into reactor 1 then enters reactor 2 via the organic phase. This leads to an increase in the amount of isobutanol so that, after separation by distillation, some of the isobutanol can be used to produce synthesis gas. Wastewater from the bottom of the extraction column is fractionated. The low-boiling compounds formed as overhead products are also used to produce synthesis gas (Fig. 25). Wastewater from the fractionating column is treated in the biological wastewater treatment plant.

3.2.1.8. Optimization of Ester Waxoil Production and Recovery of Auxiliary Products [35], [105]

Ester waxoils are produced from crude montan wax by oxidation with chromic acid to acid waxes, followed by esterification with aliphatic polyalcohols. The products are used in polishes for automobiles and furniture.

The crude montan wax contains resinous components that are removed by methylene chloride treatment. The methylene chloride is then recovered by distillation of the solution obtained (Fig. 26). In detail, the process consists of mixing the crude wax with methylene chloride (a) and filtering the mixture through a band filter (b). The CH_2Cl_2 is recovered from the filtrate (resin dissolved in CH_2Cl_2) with the aid of a recirculating evaporator (d), separating vessel (i), and condenser (e) and the residual resin can be

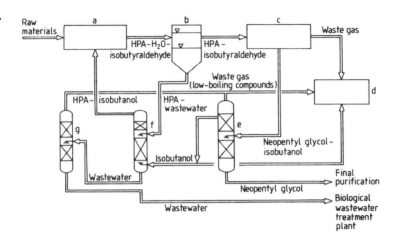

Figure 25. Production of neopentyl glycol — recovery of hydroxypivalaldehyde from wastewater by extraction with isobutanol
a) Reactor 1 (aldol condensation); b) Separating vessel; c) Reactor 2 (hydrogenation); d) Synthesis gas production; e) Distillation; f) Extraction; g) Rectification HPA = Hydroxypivalaldehyde

incinerated (f). The deresinified crude wax is melted (c) and is stripped with steam to remove any remaining CH_2Cl_2 (h).

The subsequent oxidation process with chromic acid is optimized. The earlier batchwise process in lead-lined stirred vessels resulted in contamination of water, and the consumption of raw materials and energy was high. This has now been replaced by a continuous process in which montan wax and a chromic acid–sulfuric acid mixture are passed through a sieve tray column. The two components are intensively mixed by compressed air as they enter the column (Fig. 27, a), giving a high conversion of chromic acid. The exothermic reaction heats the mixture to the temperature required, so that external heating of the column is unnecessary. This procedure gives an oxidation product of uniform quality, leads to savings in energy and raw materials, and enables the purification stage to be performed continuously and effectively. The steam saving is 1 t per tonne of wax acid compared with the previous process. The consumption of chromic acid is reduced from 170 % of the crude montan wax (former process) to 115 %. The consumption of electricity for regeneration of the exhausted chromic acid, [i.e., for electrolytic oxidation (i) of chromium(III) to chromium(VI)] is reduced by 2000 kW · h per tonne of acid wax.

The exhausted chromic acid [chromium(III) salt solution] produced in the oxidation stage is separated from the acid wax by phase separation (b) and flotation processes (c) and is electrolytically treated (i). The anolyte [chromium(VI) salt] is fed into the oxidation column (a). The catholyte [chromium(III)–chromium(II) salt mixture] is partly fed into an oxidation tower (d) and partly mixed with wastewater (spray and wash water), which contains chromium(VI). This causes the chromium(II) present in the catholyte to be oxidized and the chromium(VI) in the wastewater to be reduced. In both cases, a solution of chromium(III) is formed. The small amount of chromium in

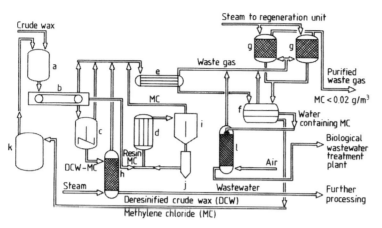

Figure 26. Methylene chloride recovery in wax production
a) Mixing vessel; b) Filter unit; c) Melting vessel; d) Evaporator; e) Condenser; f) Separating vessel; g) Activated carbon filter; h) Stripping column; i) Separator; j) Resin Incineration; k) Receiver; l) Stripping column

the wastewater is precipitated as chromium(III) hydroxide by addition of sodium hydroxide solution (f), and then filtered off in a filter press (g) and brought into solution in the acid conditions in the mixing tank (h). This is followed by the electrolytic oxidation stage to form chromic acid (i). Residual wax from the flotation stage is oxidized later in a separate operation.

Separation of chromic acid residues from the oxidation product is carried out continuously today by a disk separator. Zirconium is the only construction material that can be considered in view of the highly corrosive properties of the medium. A disk separator of zirconium has been developed specifically for this purpose in cooperation with an engineering company. This gives both an improvement in product quality by reducing the chromium content of the oxidation product and a saving of sulfuric acid for the washing processes. The sodium sulfate content of the wastewater is thus reduced by 13 kg per tonne of acid wax.

3.2.1.9. Biochemical Production of 7-Aminocephalosporanic Acid [106]–[108]

Introduction. 7-Aminocephalosporanic acid (7-ACS) is a pharmaceutical chemical obtained by elimination of α-aminoadipic acid from the fermentation product cephalosporin C. 7-Aminocephalosporanic acid is the key product for most semisynthetic cephalosporin antibiotics. It is used by Hoechst to produce Claforan (cefotaxime sodium), cefodizime, Cefquinom, and Cefpirom. Until now, 7-ACS has been produced by a chemical process. A process based on biochemical catalysts has now been developed and is considerably more environmentally friendly.

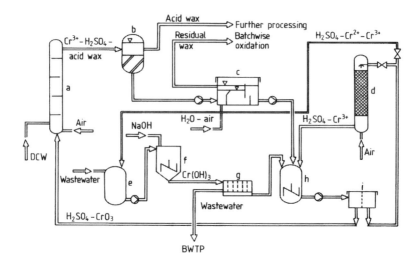

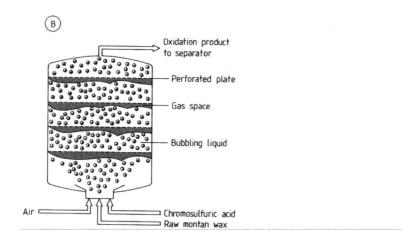

Chemical Process. The chemical process used hitherto consists of the following stages:

1) Production of the zinc salt of cephalosporin C
2) Reaction of the cephalosporin C zinc salt with trimethylchlorosilane to protect the functional groups (NH_2 and COOH)
3) Productions of the imide chloride from the silylated product by reaction with phosphorus pentachloride at 0 °C
4) Hydrolysis of the imide chloride (see Fig. 28) to yield 7-ACS.

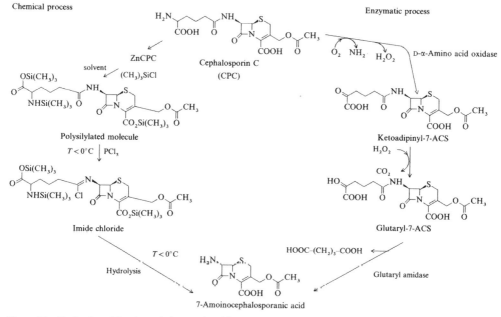

Figure 28. Production of 7-aminocephalosporanic acid

Enzymatic Process. The enzymatic synthesis is carried out as a two-stage process:

1) Cephalosporin C is supplied directly to the enzymatic stage. The side chain of the molecule is oxidatively deaminated by a D-amino acid oxidase (obtained from yeast). Oxygen is consumed, and ketoadipinyl-7-ACS, ammonia, and hydrogen peroxide are formed. Under these reaction conditions, oxidative decarboxylation to form glutaryl-7-ACS occurs spontaneously.

2) In the second stage of the process, the glutaric acid is split off by the enzyme glutaryl amidase, and 7-ACS is liberated (see Fig. 28).

The reactions are carried out at room temperature in aqueous solution. The enzymes are immobilized on a spherical carrier and can be reused many times. The enzyme D-amino acid oxidase is obtained in good yield from the natural host bacterial strain.

In the production of the enzyme glutaryl amidase, only a low concentration was produced by the wild strain *Pseudomonas*. An economically interesting concentration (greater than fiftyfold increase per volume unit) can be produced only by genetic engineering (i.e., cloning the enzyme glutaryl amidase in *Escherichia coli*).

Comparison of the two Processes. The chemical process is inferior to the enzymatic process with respect to the reactants and auxiliaries employed:

1) Use of chlorinated hydrocarbons and toxic auxiliaries
2) Use of phosphorus pentachloride, which is subject to the German decree about emergencies (Störfallverordnung), and of trimethylchlorosilane, which can be handled only in the absence of moisture and is highly flammable
3) Use of heavy-metal salts (i.e., zinc salts)

Use of the enzymatic process leads to the following changes in the amounts of residues produced (per tonne of 7-ACS) compared with the chemical process:

1) Mother liquors requiring incineration: reduction from 29 to 0.3 t
2) Waste-gas emissions: reduction from 7.5 to 1.0 kg
3) Wastewater contamination (COD): increase from 0.1 to 1.7 t
4) Residual zinc by recovery as $Zn(NH_4)PO_4$ and reutilization: reduction from 1.8 to 0 t
5) Distillation residues (from the distillative treatment of the chlorinated hydrocarbons) for incineration as waste: reduction from 2 to 0 t

Furthermore, the chemical process is energy intensive, since it can be carried out only at low temperature.

The fraction of the process costs used for environmental protection is reduced from 21 to 1%. This fraction includes the cost of the additive environmental protection (waste incineration and purification of wastewater and waste gas). Thus, the absolute environmental protection costs are reduced by 90% per tonne of 7-ACS.

3.2.1.10. Production of *n*-Valeraldehyde and Amyl alcohol

The CFC substitutes for the refrigeration sector require new lubricants for operation of the compressors. Ester of *n*-valeric acid are used for this purpose. The starting product for this acid is *n*-valeraldehyde. The production plant concept involves manufacture of *n*-valeraldehyde and amyl alcohol.

Previously (A) a mixture of 1-butene and 2-butene was converted to valeraldehyde by hydroformylation with a cobalt catalyst. Part of the aldehyde was hydrated to amyl alcohol on a nickel catalyst. The new process (B) consists of two hydroformylation stages (Fig. 29).

In the first stage, 1-butene is preferentially converted to valeraldehyde. This takes place in a low-pressure process, in which the catalyst used is a rhodium complex compound containing water-soluble phosphines as ligands. A solubilizer is also added. In the second stage, the 2-butene not converted in the first stage is hydroformylated with a cobalt catalyst. Part of the valeraldehyde obtained is then hydrogenated to amyl alcohol [109] – [111].

This modification to the process very considerably increases the yield of *n*-valeraldehyde compound and reduces the iso content. Furthermore, wastewater pollution is cut from 3.4 t COD to 1.1 t COD and the amount of waste from 785 t to 100 t (based on target product quantity of 1,000 t *n*-valeraldehyde and 1,000 t amyl alcohol) (Fig. 30).

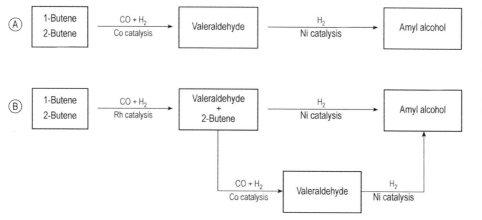

Figure 29. Production of *n*-valeraldehyde and amyl alcohol

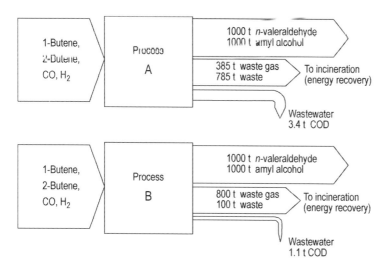

Figure 30. Comparison of previous and new production process of *n*-valeraldehyde and amyl alcohol

3.2.1.11. Production of Theobromine

Theobromine is an intermediate for the production of the vasotherapeutic agent TRENTAL. HEXTOL (Fig. 31) is a similar therapeutic agent.

Theobromine is prepared by methylation of 3-methylxanthine. In this process, theophylline and caffeine are obtained as by-products (Fig. 32).

Previously, the methylation was carried out with dimethyl sulfate in methanolic solution. The methanol in the reaction mixture was distilled off. The mother liquor resulting from the subsequent purifying operations – purification stage 1 – was also

57

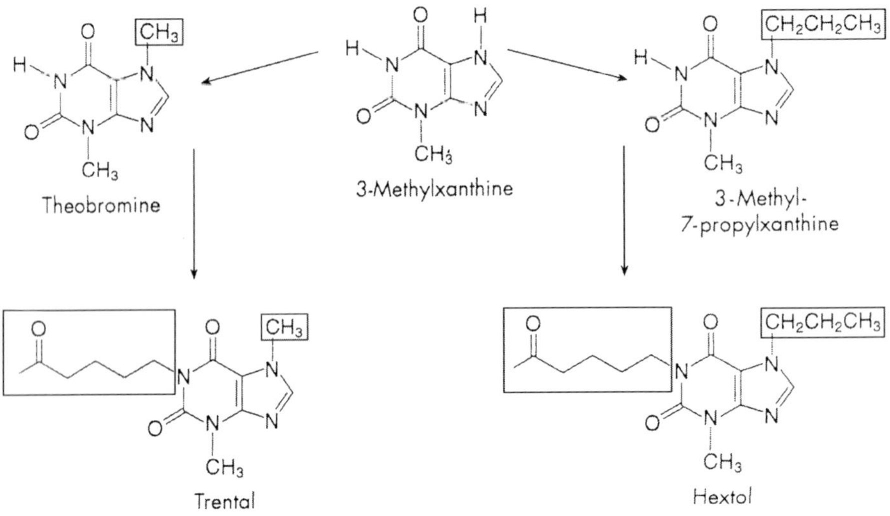

Figure 31. Manufacturing processes with 3-methylxanthine

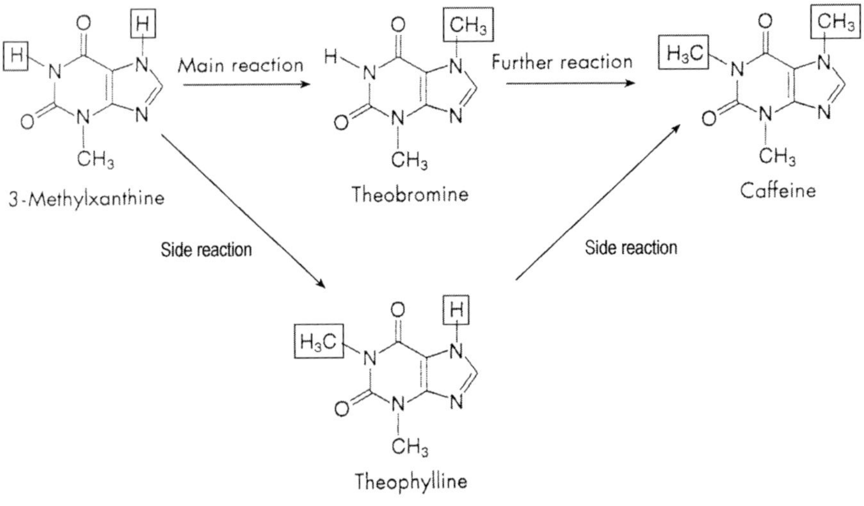

Figure 32. Production of theobromine

distilled and methanol recovered in the process. Rectification residue and mother liquor from purification stage 2 had to be discharged into the biological wastewater treatment plant (Fig. 33). Residues from filtration unit and distillation unit are combusted as waste.

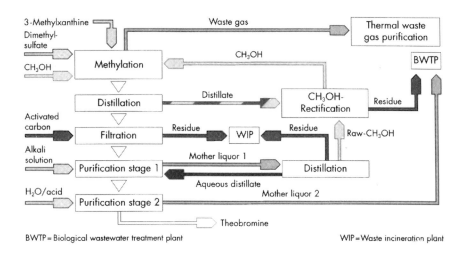

Figure 33. Production of theobromine   Previous process
BWTP = Biological wastewater treatment plant; WIP = Waste incineration plant

In the new process [112], [113] methylation is carried out with a phase transformation catalyst based on a quaternary ammonium or phosphonium compound (Fig. 34) in the present of a linear polyether in a two-phase mixture. Methanol is no longer used as the reaction medium and so energy-intensive distillation ceases to be necessary. Toxic dimethyl sulfate is replaced by methyl chloride as the methylating agent. The catalyst is recycled. Only spent catalyst with filtration residue are combusted as waste.

The mother liquor from purification stage 2 is returned to the methylation as a starting solution. Mother liquor arising from purification stage 1 that is discharged into the biological wastewater treatment plant show a 85 % reduction in DOC.

Through this process modification, the stoichiometric yield of 81 % is increased to 88 %. Calculated on 1,000 kg theobromine, the amount of waste is reduced from 7,400 kg to 50 kg. Wastewater pollution drops from 400 kg COD to 60 kg COD (Fig. 35). The energy requirement is 4 kWh (formers process: 44 kWh). Only 1,340 m^3 cooling water are needed compared with 6,300 m^3 for the previous process. Hence the wastewater generated by flue-gas scrubbing in the waste incineration plant is also reduced.

Process costs are reduced by about 15 %.

3.2.1.12. Recovery of Organic Solvents

The processing of reaction mixtures (e.g. through distillation and extraction) often involves the recovery of organic solvents. These recovered with high purity and recycled into production process. Process economy is thus improved.

There are two types of solvent recovery (Fig. 36) and they can be differentiated as follows [35]:

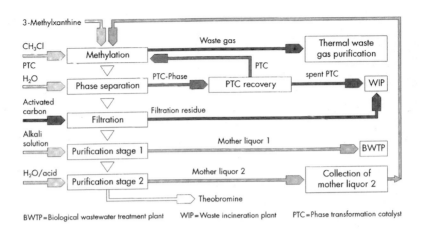

Figure 34. Production of theobromine, New process
BWTP = Biological wastewater treatment plant; WIP = Waste incineration plant; PTC = Phase transformation catalyst

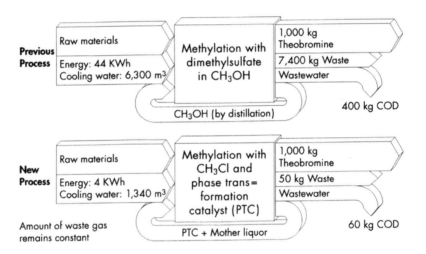

Figure 35. Comparison of previous and new production process of theobromine

- *Recovery in the main process.* The product is dissolved in an organic solvent, arising e.g. from extraction of the reaction mixture. At the end of the process the product is recovered by distillation and the solvent returned to the process.
- *Recovery in bypass.* The target product is contaminated with a by-product. The latter is separated by extraction with a solvent. The purified target product is forwarded to the next stage of treatment. The by-product, e.g. a resin remains in solution. The solvent is recovered by distillation and returned to the process. The by-product remains as residue and is treated as waste (see also Figure 26 A). Various solvent recovery

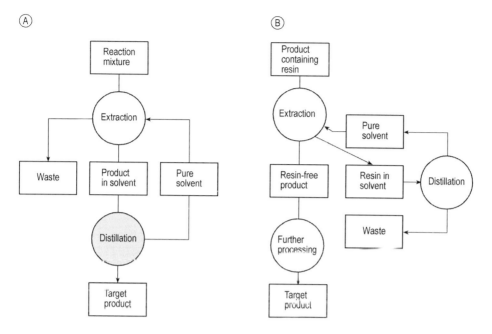

Figure 36. Solvent recovery A) In the main process; B) In bypass

a) *recovery of diisopropylether* [35]. During the preparation of recorcin and extraction of the reaction mixture is carried out recorcin is soluble in diisopropylether. The ether solution is distilled and the ether returned to the process cycle. The ether impurity in the recorcin is stripped out yielding an ether/water mixture that is led to a decanter for separation (Fig. 37).

b) *Recovery of hexanol* [67]. The production of p-nitrophenol (PNP) yields a solution (mother liquor) contraining PNP. This product is extracted with hexanol. The head product from the extraction column — a hexanol/PNP/water mixture — is distilled (1st distillation) to yield a PNP/water mixture and a hexanol/water mixture. The separated aqueous phase from separator vessel 1 and the hexanol/water fraction from the extraction column are distilled [2nd distillation (stripping column)]. Water and hexanol form an azeotrope at 97.8 °C (67 % water and 33 % hexanol). The head product from the stripping column is enriched with hexanol up to azeotropic composition. The hexanol is recovered by condensation and phase separation in separator vessel 2. (Fig. 38). The COD of the wastewater is thus reduced by 350 t/a. As a consequence the amount of sewage sludge is also reduced.

c) *Recovery of toluene* [67]. At one stage in the production of an intermediate, it is present as a solution in toluene. The toluene is recovered by distillation with a falling stream evaporator and a packed column, followed by separation of a water/toluene mixture (Fig. 39). The offgas contains toluene. This is removed by activated cartbon. The activated carbon is regenerated using stream and the solvent recovered (Fig. 40).

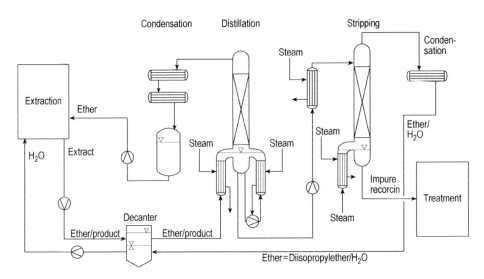

Figure 37. Recorcin production by extraction and solvent recovery

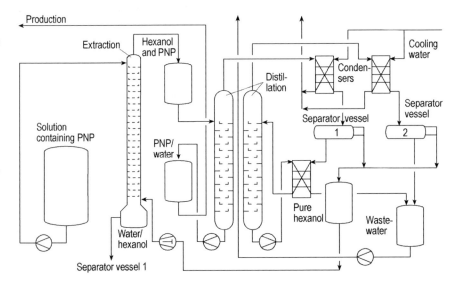

Figure 38. Hexanol recovery

d) *Recovery of benzine* [114]. Benzine is used in a special production process for a plastic. This solvent is part of 3 streams and is separated as follows (Fig. 41):
  – Liquid benzine containing plastic residues: Distillation. The plastic residues recovered from the distillation column tailings are treated as a waste product.
  – Waste gas containing benzine: The gas is cooled and fed to a activated carbon column. The activated carbon is regenerated with stream and the benzine recovered by condensation.

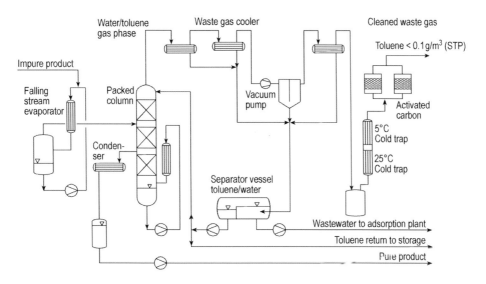

Figure 39. Toluene recovery plant

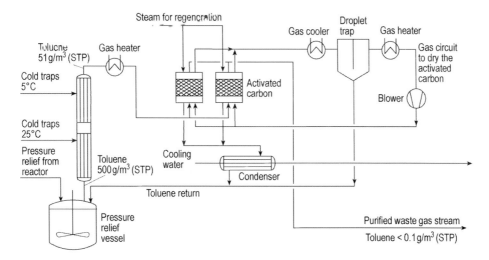

Figure 40. Toluene recovery from a waste gas purification plant

- Water/hydrocarbon mixture: Condensation, separation of the liquid water/benzine. The waste gas is ignited and the wastewater is fed to a treatment plant.
e) *Recovery of methanol and toluene* [115]. During the production of a textile chemical a solution (mother liquor) is produced containing methanol and an aromatic solvent (toluene or xylene). An azeotrope containing methanol and toluene is produced by discontinuous ditillation. The toluene-containing phase separates out on addition of water. Methanol is recovered from the methanol-containing phase by continuous

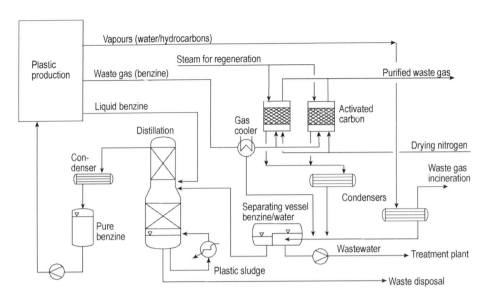

Figure 41. Benzine recovery

rectification (Fig. 42). The solvent mixture is also recovered from the waste gas by low temperature condensation (+2 to −40 °C) (Fig. 43) and separated by distillation.

f) *Recovery of ethanol and trichlorethylene* [5], [35]. The waste gas for belt dryers for ceramic components contains ethanol and trichlorethylene. These solvents are absorbed in a countercurrent washer using glycol ether as extractant. The glycol ether containing the dissolved solvents is warmed and regenerated at low pressure. Water is used to extract the solvent. The glycol ether is cooled and returned to the absorption column. The solvents are recovered as condensate. The trichlorethylene containing phase is separated and ethanol is recovered by distillation, followed by drying with molecular sieve (Fig. 44).

g) *Recovery of dimethyl formamide* [68]. During the spinning of polyacrylnitrile a mixture of dimethyl formamide (DMF) and water is produced. This mixture is separated using two distillation columns in tandem. Most of the water is removed in the first column. The mixture is separated into its components in the second column. The two columns without energy coupling are shown in Fig. 45. The plant layout was changed and the energy flows in both columns were coupled. The steam from the top of the first column is fed to a heat exchanger at the base of the second column (Fig. 46). Thus not only is DMF recovered but also because of the new plant layout the steam usage is reduced by ca. 50%. This reduction in energy usage leads to a financial saving. The disadvantage is that coupling the energy transfer between the two columns means that variation in the first column gives rise to charges in the second. Thus the demands on process control are increased. The processes are developed by use of control technology measures.

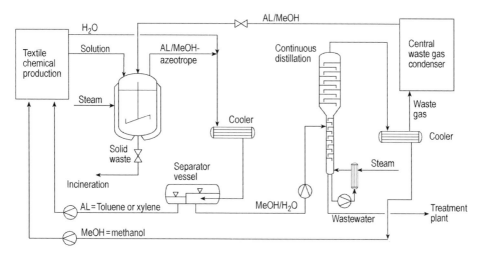

Figure 42. Recovery of Aromatic Solvents/Methanol Distillation

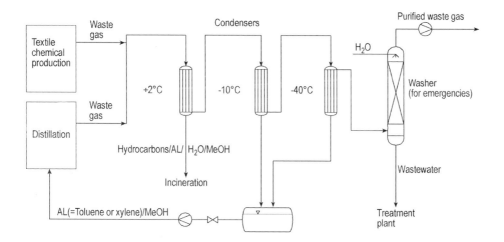

Figure 43. Recovery of Aromatic Solvents/Methanol Central waste gas condenser

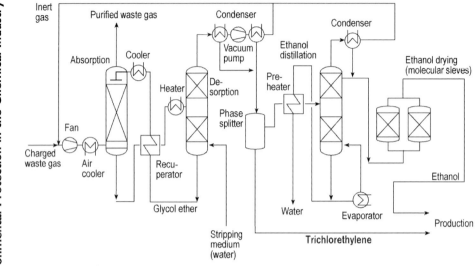

Figure 44. Trichloroethylene/ethanol recovery

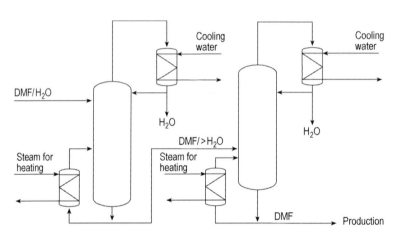

Figure 45. Recovery of dimethyl formamide (DMF) Distillation without energy transfer

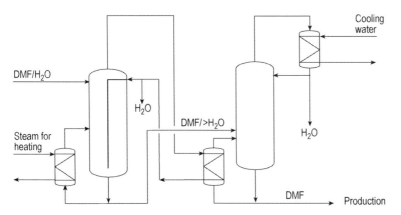

Figure 46. Recovery of dimethyl formamide (DMF) Energetically coupled distillation

3.2.2. Examples from Bayer

3.2.2.1. Avoidance of Wastewater and Residues in the Production of H Acid (1-Amino-8-hydroxy-naphthalene-3,6-disulfonic acid)

Introduction. Development of the so-called coal tar dye industry has resulted in extensive knowledge of naphthalene chemistry. The "letter acids" based on naphthalene derivatives are very important, especially for azo dyes. The color properties of the chromophore present depend highly on the substitution pattern of the naphthalene nucleus. The introduction of sulfonate, nitro, amino, and hydroxyl groups in various positions is therefore an important characteristic feature of the production of these intermediates.

A dye component that is used frequently is H acid (1-amino-8-hydroxynaphthalene-3,6-disulfonic acid). Production of this comparatively complex substitution pattern requires a series of different chemical reactions. The known side reactions and subsequent reactions on sulfonation and nitration of the naphthalene system and high-pressure hydrolysis of its substituents reduce the yield of the target product. If, as a consequence, intermediate isolations are unavoidable during the course of the process to prevent a gradual increase in the amount of by-products, additional product losses in mother and wash liquors will occur. Management of the often considerable quantities of solid waste and wastewater is therefore an important consideration in the production of H acid.

Conventional Process. The production process for H acid is divided into four basic operations each of which includes several steps:

1) Sulfonation of naphthalene to give the trisubstituted product
2) Nitration of the trisulfonic acid
3) Reduction of the nitro group to T acid
4) High-pressure hydrolysis of the sulfonic acid group in the α-position to give H acid

This long reaction chain was distributed over various buildings of the Leverkusen works.

The rule of ARMSTRONG and WYNNE applies to the multiple sulfonation of naphthalene; i.e., the two sulfonic acid groups that enter the ring are in neither the ortho, the para, nor the peri position to each other [116]. Therefore, no further reaction to form naphthalenetetrasulfonic acid occurs after the desired 1,3,6-substitution. However, this pattern can be obtained only via the 1,6- or 2,7-disulfonic acid as intermediates. The desired yield is therefore influenced considerably at the very beginning of the reaction by the selectivity and completeness of the reaction forming the disulfonic acid intermediate stage. A considerable proportion of the naphthalene used is lost at the sulfonation stage according to [117].

The sulfonation product is then nitrated. Reaction conditions can be chosen such that the α-position is preferred [118]. The directing effect of the sulfonic acid groups promotes substitution at the desired position. However, nitration of the naphthalene nucleus has a much greater tendency to lead to side reactions (e.g., oxidations) than is the case with the benzene series, so that when the nitrated mixture is diluted with water, nitrous gases are evolved.

The acid solution is neutralized by adding calcium carbonate, and the precipitated gypsum produced from the excess sulfuric acid is filtered off. The nitrated product remains in the filtrate and is reduced by iron chippings to T acid (1-aminonaphthalene-3,6,8-trisulfonic acid). The remaining calcium ions are removed as calcium carbonate together with the iron oxide by adding sodium carbonate solution, and the T acid is isolated as an intermediate by acidification of the filtrate, so that the various by-products can be removed with the mother liquor.

Naphthalenesulfonic acids are hydrolyzed to hydroxy derivatives under drastic alkaline conditions, and sulfonic acid groups in the α-position react preferentially [119].

Thus, the T acid is converted to H acid by treatment with sodium hydroxide solution under pressure [120]. The H acid is then salted out with sodium chloride. Because of the unfavorable solubility properties, large quantities of sodium chloride must be employed which remain in the mother liquor with the organic by-products.

In view of the yields of individual stages quoted in the literature, even after many years of production experience, this process gives an overall yield of only 40 % based on naphthalene.

Revised H Acid Process. The poor infrastructure (various stages of the process carried out at different locations), age of the plants, and intensity of emissions, among other factors, have provided the motivation to attempt a systematic revision of the process. This has led to complete redesign of the operation in the factory at Brunsbüttel [121], [122]. Operation in Brunsbüttel began in 1981, and the old plants have been completely replaced.

The basic process pattern is unchanged. The important changes to the process are replacement of several intermediate steps by a continuous system, and replacement of iron reduction by hydrogenation with hydrogen using a nickel catalyst [123]. A modern process control system is used that considerably reduces variation of the process control parameters compared with the earlier in situ manually controlled system. Salting out and isolation of intermediates (e.g., T acid) are omitted. There is now only the isolation of the target product. Waste-air and wastewater emissions that still occur are purified in new plants. Those waste-gas streams from the plant that contain nitrogen oxides, organic compounds, and hydrogen are treated in a combustion plant. The remaining wastewater stream containing organic by-products from the synthesis that are difficult to degrade biologically is reacted in a wet oxidation plant with oxygen at high pressure. This "wet incineration" proceeds almost to completion, so that subsequent biological wastewater purification is unnecessary.

Ecological Aspects of the Process. In comparison to the earlier process, the modern H acid production process shows considerable improvements in emission levels for equal quantities of the target product.

Results of the revision of the process
1) Starting materials
    - Naphthalene — 20 %
    - Sulfuric acid — 20 %
    - Calcium carbonate — 20 %
    - Nitric acid — 20 %
2) Gypsum — 38 %
3) Iron oxide sludge −100 %
4) Volume of wastewater − 70 %
5) Wastewater load
    - Sodium chloride −100 %
    - COD − 96 %
    - Sodium sulfate + 12 %

Efficiency. Consumption of the raw materials naphthalene, sulfuric acid in various concentrations, nitric acid, and calcium carbonate is now 20% lower.

Gypsum and Calcium Carbonate Sludge. Because of their improved quality, gypsum and calcium carbonate sludge are no longer dumped, but are supplied to the cement industry for use as raw materials. Also, the quantity of gypsum produced has been reduced by 38%.

Iron Oxide Sludge. Attempts were made to utilize the iron oxide sludge produced in the reduction by iron chippings of the nitrated compound to T acid. Use of this sludge as a raw material for the production of iron oxide pigment or in iron smelting was not very successful, so it was disposed of mainly by dumping. The present hydrogenation process produces only water as the by-product. Catalyst recovered from the process is reutilized.

Volume of Wastewater. The omission of intermediate isolation stages, the change to continuous operation of various intermediate reaction stages, and the introduction of an intermediate evaporation stage led to a reduction by 70% in the amount of process wastewater that must be treated because of its organic content.

Organic Load of Wastewater (*COD*). The aqueous processing steps, which are the rule in letter acid production, lead to the presence of side products formed from naphthalene in the process wastewater. The biological wastewater treatment is regularly unable to eliminate these materials (elimination rate sometimes <20% within normal holding periods) satisfactorily.

The efficiency of conversion of raw materials is now considerably improved, and amounts of unavoidable by-products have been reduced by high-pressure wet oxidation, so that the COD is now reduced by $\geq 96\%$. Wastewater can therefore now be discharged directly into the Elbe.

Salt Content of Wastewater. Because of the considerably smaller amounts of water used in the process, the former technique of isolating H acid by salting out could be omitted. The process no longer requires the addition of sodium chloride, so 4.7 t of sodium chloride per tonne of H acid is no longer used. However, the increase in sodium sulfate content of 12% (i.e., a total of 3.8 t per tonne of H acid) has to be accepted, but nonetheless the salt load of the wastewater is considerably reduced.

Economic Constraints. The example described is an illustration of the opportunities afforded by modern process technology to improve a production process that has remained the same in principle for 100 years. However, notwithstanding the process improvements achieved, the cost of the new plant and installations presents an increasing problem when viewed against the background of prices in international competition. Especially in Asia, obsolete technology and the associated increase in environmental pollution are tolerated. This results in cost structures that have consequences for the world market when they are widely used to confer market advantages.

The example therefore raises the question to what extent production-integrated environmental protection is hindered by lack of international harmonization. Environmental protection measures require a long-term positive outlook with respect to both market forecasts and official regulations, because the time scales for systematic process modernization, licensing, new plant construction, and an appropriate initial operating phase for refinancing must be taken into account.

3.2.2.2. High-Yield Production of Alkanesulfonates by Means of Membrane Technology [124], [125]

Introduction. Alkanesulfonates are anionic surfactants. They are produced by Bayer through the sulfochlorination process. Optimum properties are obtained when a mixture of straight-chain alkanes with 14–17 carbon atoms is used.

The alkanesulfonyl chloride (Mersol) produced by the sulfochlorination reaction is hydrolyzed with sodium hydroxide solution to give the sodium sulfonate. The hydrolyzed mixture separates on cooling into a product-rich upper phase ("glue") and a salt-containing lower phase that still contains ca. 2–3 % sulfonate and must be disposed of.

Conventional Method of Recovering the Alkanesulfonates. The reaction product of sulfochlorination (obtained in ca. 30–33% yield) is hydrolyzed at ca. 100 °C with aqueous sodium hydroxide, and the mixture obtained is cooled in several stages. Most of the unreacted alkanes separate in the hot phase and are recycled to the production process. The aqueous hydrolyzed solution contains residual alkanes, alkanesulfonates, and sodium chloride produced from reaction of the sulfonylchloride with sodium hydroxide solution.

On cooling to ca. 6 °C, an aqueous phase separates, which contains ca. 2–3 % of alkanesulfonates, 6–7 % sodium chloride, and traces of alkanes. Further cooling of the lower phase to −6 °C decreases the sulfonate content to ca. 1.5 %.

This lower phase was formerly treated in the central wastewater treatment plant. Apart from salt, it also contains organic substances that give a COD of ca. 20 g/L. Approximately 2.8 m^3 per tonne of sulfonate of the lower phase is produced resulting in a considerable COD in the wastewater.

Membrane Process for Recovery of Alkanesulfonates. Many attempts have been made over the years to reduce the wastewater load—which represents a loss of product—by a number of different methods. These include evaporation, extraction, reverse osmosis, and ultrafiltration. All of these processes have the disadvantage of high equipment cost and high energy requirements, and the space–time yield is low. The first breakthrough came with the development of new types of membrane with a definite separating efficiency and a large surface area, so-called spiral-wound modules.

Nanofiltration is best suited for successful recovery of alkanesulfonates from the lower phase. This process operates with standard spiral-wound modules at moderate pressure and gives recovery efficiencies of ≥90 %.

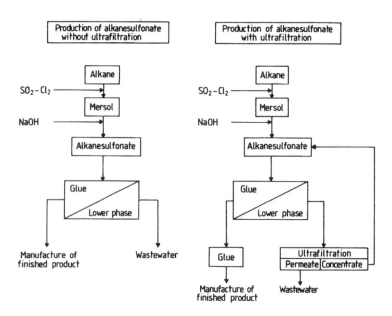

Figure 47. Comparison of processes for the production of alkanesulfonates with and without ultrafiltration

Suitable nanofiltration membranes based on the spiral-wound module technique are available from various suppliers; they usually consist of a carrier film, a spacer, and the filtration membrane. After separation, the lower phase is heated preferably to ca. 25 °C, adjusted to a predetermined pH by addition of acid, and pumped at ca. 30 bar through a battery of pressure tubes suitably arranged in parallel and fitted with spiral-wound modules.

The concentrate produced is usually recirculated until the desired concentration is reached and then returned to the production process (e.g., the hydrolysis stage).

The membrane permeate contains <10% of the initial sulfonate content and is treated in the central biological wastewater treatment plant. The entire process is operated continuously.

Surprisingly, not only do the modules have good retention properties for alkanesulfonates, but the concentrate also contains significantly less salt. The flow sheets for both processes are shown in Figure 47.

Ecological Aspects of the Modified Process. Conventional treatment of the hydrolysis mixture leads to a wastewater stream whose organic load, consisting of lost product and emulsified starting material, is ca. 20 g/L COD. The organic components also contain a certain amount of combined chlorine formed as a result of side reactions with the alkane chains during sulfochlorination.

The introduction of nanofiltration under pressure has led to a reduction in the organic content of the wastewater of up to 95%, so that the filtrate has a COD of only

0.4 g/L. This decreases the load on the factory's biological sewage treatment plant, especially that due to organically bonded chlorine.

Concluding Comments. Improvement in the efficiency of alkanesulfonate recovery was first made possible by the development of efficient multilayer membranes with long service lives and high retention properties.

A flexible system of switching between modules enables the sulfonate recovery stage to operate for long periods without cleaning or maintenance operations.

The plant was installed in 1992 in the Uerdingen factory and has fulfilled the promise of the early pilot investigations.

3.2.2.3. Selective Chlorination of Toluene in the *para*-Position

Introduction. Monochlorination of toluene— yielding usually a statistical mixture of *ortho*- and *para*-chlorotoluenes—is an important step in the introduction of functional groups into this aromatic hydrocarbon. Monochlorination represents the first stage in a large number of reaction chains leading, for example, to chlorobenzaldehydes or cresols [126]. Applications of such products include plant-protection agents, pharmaceuticals, dyes, antioxidants, and preservatives [127].

Since various product lines follow their own market laws, the demand for *ortho*- and *para*-chlorotoluene fluctuates, *ortho*-chlorotoluene often being the less desired isomer. This situation is a typical problem in joint production. If it cannot be solved by a pricing policy in accordance with market economics, the only possible solution in the most unfavorable case is to dispose of excess quantities of the less desired isomer (e.g., by incineration).

Conventional Chlorination of Toluene. In the batch chlorination of toluene (e.g., at 50 °C in the presence of $FeCl_3$ catalyst), a crude mixture is formed in which the monochlorotoluene fraction contains small amounts of *meta*-chlorotoluene, with the two isomers *ortho*- and *para*-chlorotoluene in a practically statistical distribution in the ratio *ortho* : *para* = 1.92 [128].

Since the economic importance of *para*-chlorotoluene is considerably greater than that of *ortho*-chlorotoluene, catalytic systems have been developed to increase the *para*-selectivity of the chlorination. Since the early 1930s addition of sulfur or disulfur dichloride [129] to $FeCl_3$ as a cocatalyst has been known to shift the isomer ratio in the monochlorotoluene fraction in favor of *para*-chlorotoluene (i.e., to an $o:p$ ratio of 1.11).

Growth of product lines based on *para*-chorotoluene at the end of the 1970s and early 1980s has led to further development of the cocatalyst. The use of thianthrenes [130]–[135], phenoxazines [136]–[138], or phenothiazines [139]–[142] instead of sulfur is described in patent applications.

However, these methods usually have significant disadvantages; e.g., the quantities of cocatalysts required are too large (0.1 wt% or more), the $o:p$ ratio can be kept only if antimony compounds are used as the main catalyst, or the reaction temperature is too low (20 °C or less) so that the process is uneconomical.

Exceptions to this include the use of a mixture of $FeCl_3$ and polychlorothianthrene [135] or *N*-substituted phenothiazines (e.g., $FeCl_3$ – *N*-chlorocarbonylphenothiazine) [139]. However, these more economical variations achieve at best an $o:p$ ratio of 0.84.

The market tendencies in recent years (e.g., due to new developments and other changes in the plant-protection field) are increasingly incompatible with this isomer distribution. The demand for *para*-chlorotoluene is increasing while at the same time the demand for *ortho*-chlorotoluene remains static or tends to decrease.

New Cocatalysts in Toluene Chlorination. Intensive research into the development of a catalyst system suitable for industrial use has therefore been started. These investigations have resulted in the discovery of the benzothiazepine [143]–[146] cocatalysts, whose properties are greatly superior to those of current state-of-the-art products. Their *para*-selectivity in comparison with $FeCl_3$ – sulfur ($o:p = 1.11$) is much increased (up to $o:p = 0.55$) by using minimum quantities of cocatalyst (e.g., 30–50 ppm) in the presence of $FeCl_3$ as the main catalyst and at industrially favorable temperatures of 40–50 °C.

After a relatively short optimization phase to establish the exact composition of the catalyst system, the reaction temperature, and the exact flow rates, etc., the new system could be used in an existing continuous full-scale plant.

Hence, in principle, since the early 1990s, the isomeric ratio in toluene chlorination has been adjustable within the range $o:p = 1.11$ to $o:p = 0.55$ according to market conditions; i.e., for each tonne of *para*-chlorotoluene produced, the amount of *ortho*-chlorotoluene can be fixed within the range 1.11 to 0.55 t. A comparison of old and new processes is given in Figure 48.

Ecological Aspects of the Catalyst Change. The decreasing demand for *ortho*-chlorotoluene has limited the possibilities for utilizing the excess material. The effect on cost that would result from incineration of *ortho*-chlorotoluene had to be assessed. With this background, the development of new benzothiazepine cocatalysts relieved the isomer problem at just the right time. The isomer ratio has now been moved in the desired direction and adapted to the demand pattern. No incineration of an undesirable product is necessary.

Old process:

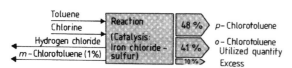

New process:

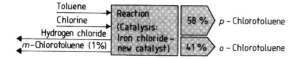

Figure 48. Selective chlorination of toluene in the *para*-position — comparison of old and new process

Concluding Comments. Extensive fundamental research into further development of catalysts has been supported by the use of theoretical chemistry and modern mathematical methods based on random techniques. The plants in which these results have been realized since 1989 operate in the Leverkusen works.

3.2.2.4. Production of Naphthalenedisulfonic Acid with Closed Recycling of Auxiliaries

General. Sulfonation of the naphthalene ring system has a long tradition, particularly as an important process step in the synthesis of the numerous "letter acids," which are important intermediates in dye production. The process parameters (e.g., exclusion of water and temperature control) determine the substitution pattern (positions 1 or 2 in the naphthalene ring system). With an appropriate excess of sulfonating agent, both rings of naphthalene can be substituted, or more highly sulfonated derivatives can be obtained. Since undesired isomers and by-products are formed, sometimes in considerable proportions, the process steps for workup and product isolation are frequently complex. Product yields relative to the naphthalene used, therefore, turn out to be relatively low.

Conventional Process for Production of Naphthalenedisulfonic Acid. The standard process for sulfonating aromatic hydrocarbons uses sulfuric acid and oleum as reactants. Under these conditions, the reaction vessels can be made out of gray cast iron, and steel can be used for piping and fittings.

$$C_{10}H_8 + 2SO_3 \xrightarrow{H_2SO_4} C_{10}H_8O_6S_2$$

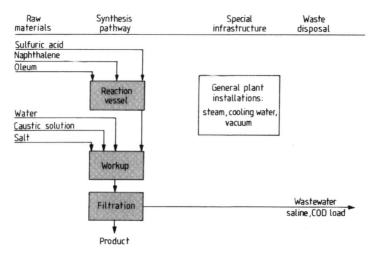

Figure 49. Conventional process for production of naphthalenedisulfonic acid

Naphthalene is introduced into sulfuric acid and reacted with added oleum [147]. A temperature of 50–60 °C is required for satisfactory space–time yields. The reaction mixture is then added to water in vessels with acid-resistant brick lining. As a result of adding sodium hydroxide solution and sodium sulfate, naphthalenedisulfonic acid precipitates as its disodium salt and can be filtered off. Aside from residual amounts of the target product, considerable proportions of undesired isomeric di- and trisulfonic acid derivatives remain in the filtrate as by-products. Because of the high saline contamination, a technically satisfactory solution to the problem of obtaining fresh sulfuric acid from these wastewater streams containing sulfuric acid has not yet been found. Likewise, current processes for biological wastewater treatment show only very mediocre elimination performances for this filtrate, so that the effluent of the wastewater treatment plant has high residual organic contamination (see Figs. 49 and 51).

New Process for Naphthalene Sulfonation. Modification of the process aims simultaneously at two goals: increasing the yield with more selective reaction and recycling sulfuric acid from the wastewater. Accordingly, not only must the parameters of the reaction step be changed so that fewer by-products are formed, but the product must be isolated by a salt-free alternative process to enable sulfuric acid to be recovered.

The desired higher selectivity of the sulfonation reaction cannot be achieved by using sulfuric acid as a reaction medium but rather by dissolving the reactants, naphthalene and sulfur trioxide, in an organic solvent (chlorinated hydrocarbon, CHC) and reacting them at temperatures lower than those of the previous process [148].

Nevertheless, because of the dehydrating action of sulfur trioxide, higher-molecular structures are formed via anhydride bridges (Scheme 1). The anhydride bridges must, therefore, be hydrolyzed in the next step. For this purpose, the reaction mixture is

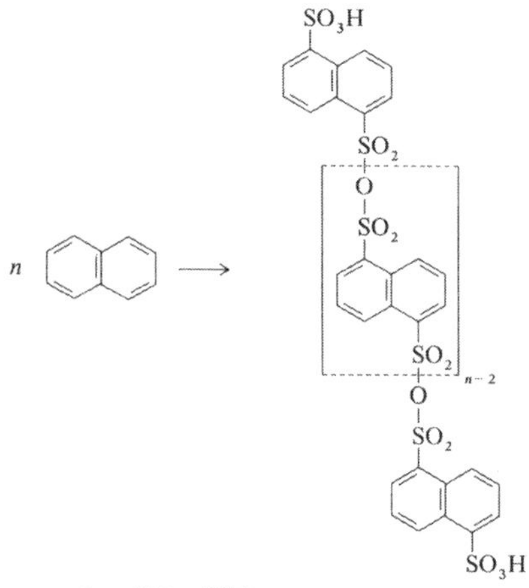

$$nC_{10}H_8 \xrightarrow[-(n-1)H_2SO_4]{+(3n-1)SO_3,\ CHC} C_{10}H_7O_5S_2(C_{10}H_6O_5S_2)_{n-2}C_{10}H_7O_6S_2$$

$$\xrightarrow{(n-1)H_2O} nC_{10}H_8O_6S_2$$

Scheme 1. Naphthalene sulfonation

added to water. An aqueous sulfonic acid solution is formed and the solvent is subsequently separated.

Subsequently, instead of the sodium salt, the free sulfonic acid [149] is precipitated by addition of sulfuric acid or recycled mother liquor, followed by cooling. The mother liquor, which remains after filtering off the product, can be supplied directly to existing plants for regeneration of spent sulfuric acid.

Ecological Aspects of the Process Alteration. The previous process is essentially characterized by a linear material flow to obtain the target product coupled with a flow of highly contaminated wastewater that contains the high naphthalene losses due to the formation of by-products, excess sulfuric acid, and inorganic salt. Two recycled streams are now superimposed on this process [150] (see Fig. 50).

The organic solvent is recovered for reuse in the production plant. Excess sulfuric acid mother liquor is conveyed via the internal piping system for spent acid to the existing sulfuric acid plant, where concentration, decomposition to sulfur dioxide, and oxidation to sulfur trioxide occur. Sulfur trioxide is then withdrawn and fed to the sulfonation reaction through piping.

In the new process, process wastewater is, therefore, collected as the distillate from spent acid concentration. Since the volatilized components from evaporation are negligible, the previous wastewater contamination of more than 3000 t/a COD, which exhibits rather unsatisfactory biological degradation, is reduced to a few tonnes per year (e.g., from shower, rinsing, or cleaning water). Thus, for this process, a COD reduction of nearly 99.9 % is achieved (see Fig. 51). Simultaneous alteration of the

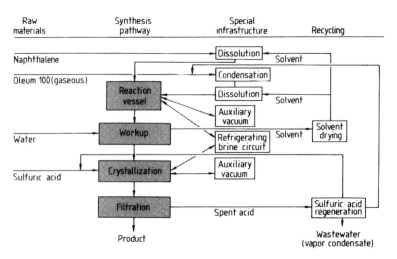

Figure 50. New process for the production of naphthalenedisulfonic acid

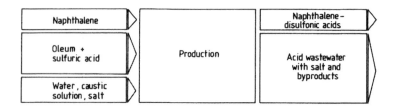

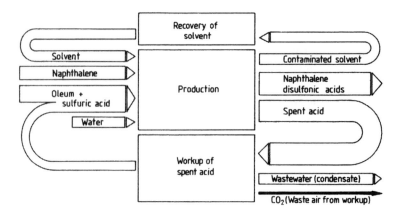

Figure 51. Mass balance of naphthalenedisulfonic acid production

reaction process markedly improves the yield, so that the proportion of naphthalene that does not react to the target product is now reduced by about a half.

Selection of the solvent however, leads, to considerable additional measures in process design and construction engineering. With regard to solvent power, enthalpy of evaporation, phase separation behavior, etc., low-boiling chlorinated hydrocarbons are best. This, for instance, involves special measures for sealing the tank farm on the water side, as well as monitoring the platform and surface drainage with detectors, so that if leaks occur, an emergency tank can be switched in.

The product is crystallized with vacuum cooling. The associated stripping effect allows maximum recycling of the solvent. Because the solvent must be anhydrous for reuse, the collected solvent streams are dried with sulfuric acid in a pulsed column. Waste air collected from the entire apparatus is supplied to a waste-air incineration plant with downstream washers.

Supplementary Engineering Equipment for Process Rearrangement. The conventional process can, in practice, be run manually with on-site operation. In comparison, the additional infrastructure of the new process with its multiple interconnections, as well as accurate maintenance of reaction parameters, requires considerable supplementary installations. Operation of the process is supported by a new control system developed specifically for this purpose. A high degree of standardization and uniform quality is achieved [151]. The new plant has been in operation in Leverkusen since 1989.

3.2.2.5. Avoiding Residues in Dye Production by Using Membrane Processes [152]–[157]

Introduction. In the production of water-soluble dyes, recovery of the product usually occurs by salting out, filtering on filter presses, redissolving or resuspending and refiltering, standardization, and drying. In this process, large quantities of wastewater and salt are produced. With the introduction of the new technique of pressure permeation, the amounts of salt and organic materials produced are reduced considerably, and a higher yield is obtained.

Conventional Process. The end product of many dye syntheses is an aqueous salt-containing dye solution. Separation of the product dye from by-products such as salts and residual starting materials is performed conventionally by addition of further large quantities of salt to the dye solution to salt out the product dye, which is then filtered off from the salt solution by means of filter presses. Liquor from the presses contains mainly the salts used in the process together with up to 5% of the product (Fig. 52).

New Process (Fig. 53). An effective new method for reducing wastewater contamination uses pressure permeation. With suitable dyes, the salting-out process and

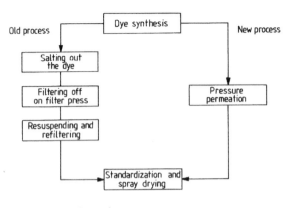

Figure 52. Flow diagrams of the old and new processes for production of a powdered dye

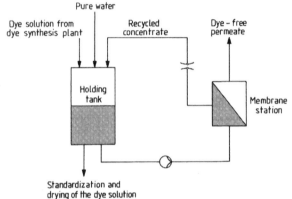

Figure 53. Flow diagram of a pressure permeation plant

isolation of dyes is replaced by salt removal and concentration of the reaction solution via membranes. The concentrated solution can then be processed further to give a directly marketable product.

Pressure permeation plants use semipermeable membranes that are permeable to water, inorganic salts, and small organic molecules, but quantitatively retain dyes in solution. According to the separating characteristics of the membrane and the resulting operating pressures to be used, separation processes are classified as microfiltration, ultrafiltration, nanofiltration, and reverse osmosis. The membrane surfaces are in the shape of spirally wound, plate-shaped, or tubular modules. The salt-containing synthesis solution passes from the reaction vessel into a holding tank and subsequently through the membranes under pressure. There it separates into a salt-containing, almost dye-free permeate, and the dye concentrate. The dye concentrate is recycled to the holding tank. After further treatment, the permeate can be reused as feedwater or otherwise disposed of as wastewater. Compared with the filtrate of the conventional process the permeate produced from the pressure permeation plant has a COD that is decreased by 80 % and a salt content (only reaction salt) decreased by 90 %. The filtration, washing, and redissolving steps, which are complex and labor intensive, are replaced by a fully automatic pressure permeation stage. Since retention of the dye

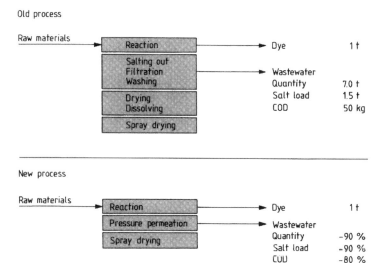

Figure 54. Residue minimization by pressure permeation

by the membrane is almost quantitative, the dye yield is increased up to 5 % by using the new pressure permeation technique.

Ecological and Economic Aspects of the Process Change. By use of the pressure permeation technique, wastewater contamination can be decreased considerably and thus the following wastewater treatment simplified. Also, the almost complete removal of salts from dyes marketed in the form of highly concentrated solutions virtually eliminates the necessity of adding auxiliaries for standardization.

In addition, the economics are improved by reducing wastewater costs, while product quality remains the same or better, and the yield is sometimes higher. The application of pressure permeation to other fields is being investigated and developed:

1) Extension of the process to other applications (dye grades and intermediate products)
2) Optimization of the synthesis and production processes (Fig. 54)

To sum up, introduction of the pressure permeation technique has the following economic advantages: (1) increase in yield, and (2) reduction of wastewater treatment costs; and hence an improved profitability with equal or better product quality.

3.2.2.6. Fuel Replacement in Sewage Sludge Combustion by Utilization of Chlorinated Hydrocarbon Side Products [158]–[161]

Introduction. In the chemical industry, much effort is devoted to preventing residue formation completely. Although in recent years, great success has been achieved in the area of residue prevention, residue formation cannot be completely avoided in the chemical industry. Attention must therefore be directed toward the most feasible methods of utilizing these residues.

Old Processes. *Production of Sewage Sludge.* In the biological wastewater treatment plants of the Leverkusen works, the organic content is degraded by microorganisms in biological stages. In this process, sewage sludge is formed: first a preliminary sediment by chemical-mechanical purification, and then excess activated sludge from the biological wastewater purification stage.

This sludge has until now been treated by thickening, conditioning, and dewatering, and been disposed of as solid filter cake on the works-owned special waste dumps.

Production of Chlorinated Hydrocarbons (CHC). During the production of, e.g., rubber chemicals, quantities of highly chlorinated hydrocarbons result as by-products, thus far unavoidably. These materials could not be disposed of at facilities on land and were incinerated in special ships on the high seas.

New Methods. *Incineration of Sewage Sludge.* To dispense with disposal of the organic materials in sewage sludge on the works-owned dump, to minimize the amount of water dumped in the sewage sludge filter cake (which contains 60% water), and to utilize the available dumping space more efficiently, alternative methods of disposing of the sludge were investigated. The following criteria were formulated for comparing the alternative methods:

1) Volume reduction
2) Process safety
3) Environmental compatibility of the residues
4) Cost

Based on these decision criteria, the method of choice for disposing of sewage sludge was incineration.

Incineration of Chlorinated Hydrocarbons. Finding a way of integrating the incineration of highly chlorinated hydrocarbons into the sludge disposal concept was very important. From the political discussion generated by incineration of these liquid residues on the high seas, one expects only incineration on land to be permitted in the future.

Description of the Process (see Fig. 55). The sewage sludge incineration plant, which has a capacity of 80 000 t/a, is divided into the following sections:

1) Multiple-hearth furnace
2) Afterburner chamber
3) Waste-heat boiler for waste-heat utilization
4) Multistage flue gas purification

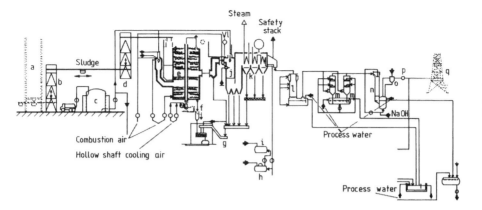

Figure 55. Schematic of sludge incineration
a) Sludge bunker; b) Elevator 1; c) CHC storage tank; d) Elevator 2; e) Multideck furnace; f) Ash removal; g) Transport container; h) Boiler feedwater tank; i) Condensate tank; j) Afterburner chamber; k) Waste-gas cooler, l) Quench tank; m) Rotary scrubber; n) Jet scrubber; o) Demister; p) Induction fan; q) Stack

The sewage sludge is conditioned; i.e., filter aids are added to convert it into a form with good filtering properties. The sludge is dewatered in chamber filter presses, stored temporarily in a bunker (a), and charged via trough feeders into the buckets of a transport plant (b, d). These buckets contain ca. 1.5 t of sludge. The sludge is charged by the buckets into a storage vessel on the top deck of the incinerator and is fed continuously from there into the furnace (e). A feed rate of up to 12 t/h sewage sludge can be achieved.

The cylindrical furnace consists of a steel casing with an interior lining and exterior insulation; it is divided into eight horizontal decks. The diameter of the furnace is 8 m, and its height 12 m.

In the center of the furnace is a vertical hollow rotating shaft with four rabble arms for each deck carrying scraper blades arranged at an angle. The rotary motion of the shaft causes the sewage sludge fed to the top deck to be swept from the inside to the outside, and on the next deck from the outside to the inside, falling through openings at the outer and inner edges of the decks to the deck below. The ash, which is produced on the two lowest decks, moves in the same way. The multideck furnace operates in countercurrent; i.e., the combustion air and hot flue gases pass through the furnace in an upward direction, drying the sludge, while the feed material is transported in the opposite direction. Also, the combustion air cools the ash, the air being thereby preheated, and the hot flue gases dry the sludge to be incinerated.

In this method of operation, three different zones are produced in the multideck furnace. From top to bottom, these are

1) Drying zone from deck 1 to deck 4, temperature 400–700 °C
2) Incineration zone in decks 5 and 6, temperature 800–1000 °C
3) Cooling zone in decks 7 and 8, 250–350 °C

At the side of the sixth deck is an uncooled combustion chamber lined with refractory materials. This is used to provide intense heat for starting up the furnace and, if required, can also be used for supplementary heating during incineration of the sludge.

Exhaust gases from the combustion zone and vapors from the drying zone of the multideck furnace pass into a lined afterburner chamber (j). Liquid organic waste (e.g., chlorinated hydrocarbons from production processes) is used as fuel in the afterburning chamber. To avoid chloride corrosion, the outer wall of the afterburner chamber is connected to the recycling system of the process gas cooler and thereby cooled. The completely burned hot waste gas then passes into the waste-gas cooler (k) where its sensible heat energy is used to produce steam. Combustion gases leave the waste-gas cooler at 250–300 °C and are then led to flue gas scrubbing. The ash produced is removed via a rotary valve (f) and fed through a chain trough conveyor (g) to the ash removal system of the multiple-hearth furnace for cooling and agglomeration. The ash produced is disposed of on the works-owned dump.

The hot flue gases at ca. 250 °C are cooled to 70–80 °C in the quench tank (l) by spraying with water, and the gaseous substances (e.g., hydrogen chloride and hydrogen fluoride) and dust particles are washed out. Any remaining fine dust and hydrochloric acid are removed from the flue gas by a two-stage rotary scrubber (m).

The flue gas is then treated with sodium hydroxide solution in a jet scrubber (n) to remove sulfur dioxide. The purified flue gas passes through a demister (o) and an induced draft fan (p), and into the stack (q).

A comparison of the flow sheets of the old and new processes is given in Figure 56.

Outlook. More than 50 % of the sewage sludge produced can now be incinerated. It is expected that all sewage sludge will be disposed of in this way by 1999.

3.2.3. Examples from BASF

3.2.3.1. Emission Reduction in Industrial Power Plants at Chemical Plant Sites by Means of Optimized Cogeneration [162]–[164]

Industrial power stations represent some of the largest individual emitters of a number of pollutants from production sites. The most important of these pollutants are CO_2, CO, SO_2, NO_x, and dust.

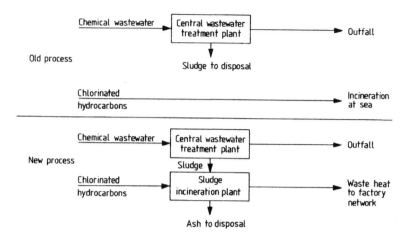

Figure 56. Comparison of old and new wastewater sludge treatment and chlorinated-hydrocarbon incineration processes

The 1972 figures for the BASF Ludwigshafen site, which contains more than 300 production plants, are given below:

Total emissions	113 000 t/a
From energy generation	55 000 t/a
From chemical production	58 000 t/a
Individual emissions	
CO	30 000 t/a
NO_x	30 000 t/a
NH_3	2 000 t/a
C_5H_{12}	2 300 t/a
HCl	500 t/a
Dust	6 300 t/a

The primary task of industrial power stations is to generate heat energy for the production processes. On large sites, this heat energy has customarily been produced in power–heat cogeneration plants, so that part of the electrical energy requirement is also produced on-site. Today, these cogeneration plants are mainly steam power plants with characteristic values of $0.15-0.25$ MW $\cdot h_{el}/(MW \cdot h_{th})$. These characteristic values on the producer side must be compared with those on the consumer side. The following values are typical:

Sites producing their own basic chemicals	$0.4-0.8$ MW $\cdot h_{el}/(MW \cdot h_{th})$
Other sites	$0.25-0.35$ MW $\cdot h_{el}/(MW \cdot h_{th})$

Comparison of these figures shows that cogeneration plants using steam-based processes cannot supply the electricity requirements of the consumer on the basis of power–heat cogeneration. On the other hand, small condensation power plants on industrial sites are generally not competitive with the large installations of electricity supply companies. This follows from the decline in capacity of industry-based power stations in West Germany (Fig. 57), which is due entirely to the closure of condensation

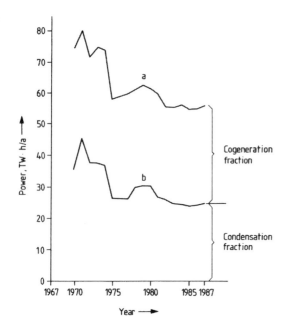

Figure 57. Breakdown of industrial electricity generation by prime mover type
a) Back pressure, tapped back pressure, gas turbines, piston engines; b) Condensation machines

plants in the industrial area. The shortfall in electrical energy is made up for by purchasing.

Several methods are used to reduce emissions from industrial power stations:

1) Lowering the energy consumption of production plants
2) Removing pollutants from flue gases of industrial power stations
3) Using alternative fuels
4) Using alternative technologies for energy conversion

An economic and ecological optimum is achieved only by a "tailored" combination of all four methods on any individual site.

Lowering Energy Consumption. Attempts to reduce energy consumption of production plants have been very successful in past years. Here, the pinch analysis method of LINNHOFF has proved very effective.

Pinch analysis is a method of assessing the works infrastructure of production plants and supply systems with respect to energy consumption, quantities of waste products, quantities of emissions, and capital costs. The fundamental elements of pinch analysis are the composite curves, the grid diagram, and the pinch itself.

In the composite curves (Fig. 58), the cold and hot streams of a production plant are plotted against temperature. When the two curves are shifted horizontally relative to each other until the minimum temperature difference necessary for heat-transfer processes is reached at the point where the curves come closest together (pinch), this gives the theoretical minimum for the heat streams that must be supplied or removed (hot and cold utility targets).

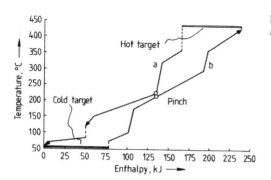

Figure 58. Hot and cold composite curves
a) Hot curve; b) Cold curve

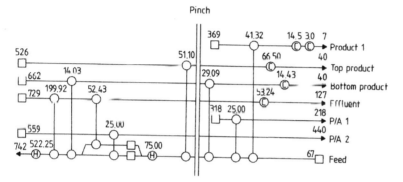

Figure 59. Grid diagram

The pinch represents the boundary between the balanced heat sources (system below the pinch) and heat sinks (system above the pinch). In the optimum case, no heat transfer occurs across the pinch. The grid diagram (Fig. 59) is used in the development of economical heat recovery systems using the pinch technology.

The result of systematic assessment of the BASF Ludwigshafen site is shown in Figure 60. The amount of sales products increased by 143 % from 1972 to 1992; however, the factory steam consumption decreased slightly. In 1992, only 33 % of the steam requirement was supplied from primary energy. Whereas the fraction of steam produced by combustion of residues remains the same, the largest fraction today is supplied by the utilization of waste heat from chemical plants. Consequently the consumption of primary energy dropped drastically.

Emission Reduction. The reduction in emissions from power plants was correspondingly drastic, and the more than proportional decrease in the use of the primary energy carrier fuel oil has had an additional effect.

At chemical sites producing sulfuric acid, further reduction in SO_2 emissions of power plants is possible using the Wellman-Lord process in which pure SO_2 is recovered and recycled to the production process. The power plant thus supplies starting materials to the chemical production plant, and becomes an integral part of it. In the

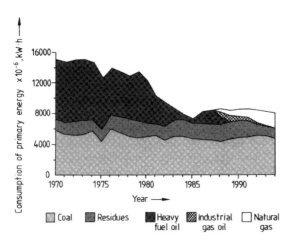

Figure 60. Consumption of primary energy at the Ludwigshafen works

Table 3. Atmospheric emissions at the Ludwigshafen site in 1992 (1996)

	Power plants, t/a	Total, t/a
CO	160	12 500 (10 100)
NO_x	240	6 000 (4 400)
SO_2	1350	2 300 (2 500)
NH_3	0	600 (380)
C_5H_{12}	0	100
HCl	26	140 (40)
Dust	200	900 (600)
Total emission	3500	24 000 (19 300)

Wellman–Lord process, SO_2 is absorbed in aqueous sodium sulfite solution with formation of sodium bisulfite:

$$SO_2 + Na_2SO_3 + H_2O \rightarrow 2\,NaHSO_3$$

The SO_2 is then desorbed by heat in a later stage.

Other additive purification processes carried out on exhaust gases led to the decrease in atmospheric emissions at the Ludwigshafen site shown in Figure 61. Emissions of individual components for the year 1992 (1996 [165]) are shown in Table 3.

Use of Alternative Technologies for Energy Conversion. A new cogeneration technology using natural gas as the energy carrier has also led to a reduction in CO_2 emission. Figure 62 shows the principal arrangements of a conventional steam turbine power plant (A), a gas turbine power plant (B), and a combined gas and steam turbine power plant (C). The range of characteristic values [$MW \cdot h_{el}/(MW \cdot h_{th})$] is also shown in each case. Industrial power stations are fundamentally heat controlled. In a modern cogeneration process using gas turbines in this way ca. three to four times the amount of electrical energy is produced compared with the steam turbine process.

So production plants can meet their total demand of electric power and even sell a considerable amount to the public. This also leads to an overall reduction of emissions.

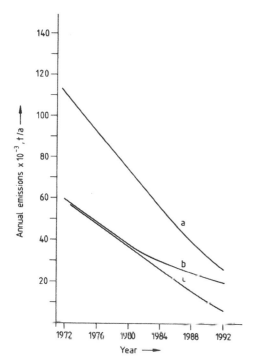

Figure 61. Atmospheric emissions from the BASF Ludwigshafen works (excluding CO_2)
a) Total emissions; b) Emissions from power plants; c) Emissions from production plants (see also Fig. 3)

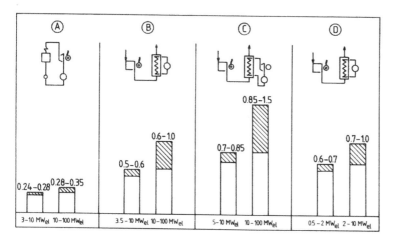

Figure 62. Characteristic values for typical cogeneration processes
A) Combined heating and power stations with steam turbines; B) Combined heating and power stations with gas turbines; C) Combination plant (without supplementary firing); D) Block-type thermal power station with gas engines

3.2.3.2. Closed-Cycle Wittig Reaction [166]–[169]

The Wittig reaction is widely used in the production of vitamins, carotenoids, pharmaceuticals, and antibiotics. It proceeds, e.g., according to the following scheme:

$$R^1-CH_2-Cl + P(C_6H_5)_3 \rightarrow R^1-CH_2-P^+(C_6H_5)_3Cl^-$$

$$R^1-CH_2-P^+(C_6H_5)_3Cl^- \xrightarrow[-HCl]{Base} R^1-CH = P(C_6H_5)_3$$

$$R^1-CH = P(C_6H_5)_3 + R^2-CHO$$
$$\rightarrow R^1-CH = CH-R^2 + O = P(C_6H_5)_3$$

The Wittig reaction is characterized by mild reaction conditions and high yields. It therefore meets the criteria for clean (smooth) chemistry. However, its more widespread use is hindered by the fact that equimolar amounts of triphenylphosphine (TPP) must be used, and the inactive triphenylphosphine oxide (TPPO) formed in the reaction is today still disposed of as waste.

The active triphenylphosphine is synthesized according to the overall equation

$$PCl_3 + 3\,C_6H_5Cl + 6\,Na \rightarrow P(C_6H_5)_3 + 6\,NaCl$$

Since TPPO is not easily degraded and phosphorus must not be present in wastewater, the only possible disposal route is as follows (Fig. 63). The TPPO-containing residues are incinerated to form P_2O_5, which is removed in a scrubber with the aid of

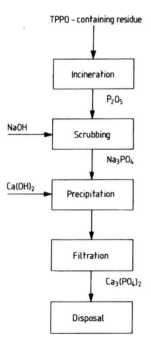

Figure 63. Flow sheet for disposal of TPPO

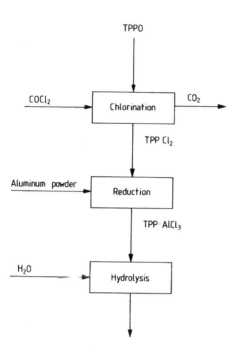

Figure 64. Flow sheet for treatment of TPPO

dilute sodium hydroxide solution. Phosphorus is then precipitated as calcium phosphate and filtered off, and the filter cake is disposed of on a hazardous waste dump. This disposal method has the following disadvantages:

1) The filter cake uses valuable dumping space
2) The P_2O_5 aerosols have a poisoning effect on the DENOX catalysts of the incineration plant
3) The P_2O_5 aerosols cause clogging of cloth filters

Attempts were therefore made to convert inactive TPPO back to active TPP, which could then be used again in the Wittig reaction. A route along the following lines was ultimately found to be successful (Fig. 64):

The TPPO-containing residues are dissolved in chlorobenzene and reacted with phosgene gas to form triphenylphosphine chloride, with liberation of CO_2. In the second step, finely divided aluminum powder is added with mixing. The TPP · $AlCl_3$ complex formed is then decomposed hydrolytically in a third reaction step, giving the desired TPP.

A flow diagram of the process is shown in Figure 65. The TPPO-containing residues from various operations are collected. Residues from other companies can also be included at this point. Pure TPPO is recovered by distillation and dissolved in chlorobenzene.

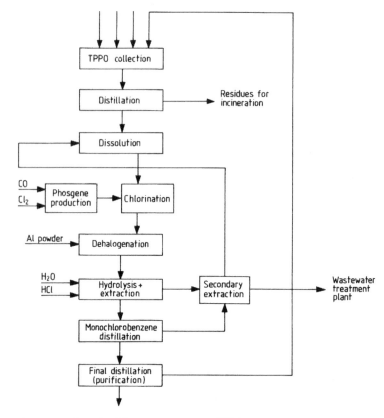

Figure 65. Schematic of the treatment process of TPPO

Chlorination is carried out in a stirred-tank reactor inside a gastight chamber. Phosgene is also generated in this chamber and fed immediately to the reactor.

Dehalogenation with aluminum powder is carried out in the next stirred-tank reactor.

The TPP · $AlCl_3$ complex is then hydrolytically decomposed with water in another stirred tank in which the pH is adjusted with HCl so that water-soluble $Al(OH)Cl_2$ is formed. The TPP is dissolved in the chlorobenzene layer. Traces of TPP are recovered from the aqueous phase by extraction with chlorobenzene. In a subsequent distillation stage, chlorobenzene is distilled off and recycled. The TPP is then purified by distillation. The aqueous phase from the extraction stage is fed to a sewage treatment plant, which operates in the weakly alkaline range, so that aluminum is precipitated as $Al(OH)_3$ and acts as a flocculation agent in the desired manner.

The ecological advantages of the new process can be seen from a comparison of the ecological balances (see Fig. 66 and 67):

This comparison shows that the new process results in the complete absence of phosphorus emissions, a reduction of the chloride load of the wastewater to one-third of its former value, and a reduction of CO_2 emission to only ca. 5 % of its former value.

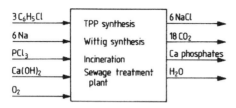

Figure 66. Ecological balance of the old process

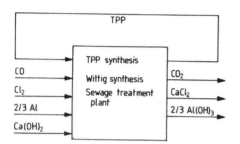

Figure 67. Ecological balance of the new process

The new low-emission process for the production of TPP enables the Wittig reaction to be used much more widely in industrial syntheses.

3.2.4. Integrated Environmental Protection and Energy Saving in the Production of Vinyl Chloride (Example from Wacker Chemie)

Introduction. The production of vinyl chloride monomer (VCM), starting material for poly(vinyl chloride) (PVC) has a long tradition. Raw materials for VCM production are ethylene, chlorine, air (oxygen), and sometimes external hydrogen chloride (HCl) from other chlorination processes. The main production steps are shown in Figure 68. This integrated process contains two internal material cycles, namely, the closed-loop recycling of nonconverted 1,2-dichloroethane (EDC) from the pyrolysis step (so-called recycle EDC) and the closed- or partly open-loop recycling of HCl formed in the cracking step back to the oxychlorination unit. With totally closed-loop recycling of the crack HCl—or if HCl must be exported due to quality reasons or the exported quantity made up by HCl from external sources—the combined VCM process runs "HCl balanced." However, if crack HCl is exported without compensation, the combined VCM process becomes "HCl unbalanced."

Differences in the processes used by individual VCM producers are due mainly to the different technologies applied in direct chlorination; these include high- or low-temperature direct chlorination (both processes are operated in the homogeneous liquid phase in the presence of oxygen as substitution inhibitor), and oxychlorination (fixed- or fluidized-bed process in the heterogeneous gas phase). Differences in these

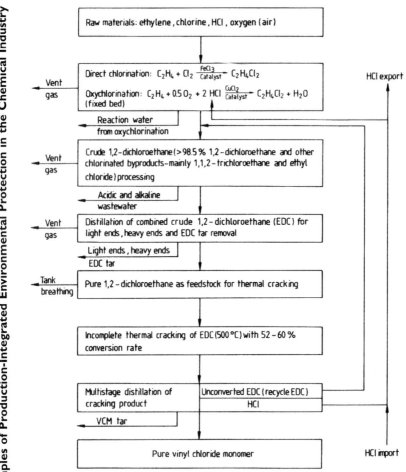

Figure 68. Main steps in production of vinyl chloride monomer

steps determine the energy requirements and the design of equipment for processing and purifying the intermediate products. The latter is as complex as pure product isolation because the reaction products formed in EDC synthesis and EDC cracking consist of multicomponent mixtures.

Conventional Production Process. Figure 69 shows the overall process flow and mass balance of the old process in operation since mid-1976. Target products in addition to VCM are hydrogen chloride for silicon chemistry and 20% hydrochloric acid for internal site demand. The combined VCM process was nearly "HCl balanced" because the HCl flow exported to silicon chemistry was balanced by imported HCl from a chlorolysis process (tetrachloroethylene production plant). The amount of by-products was all in all 4.4 wt% related to VCM production. According to JENSEN et al. [170], ca. 64% of these residues are heavy ends and EDC–VCM tars which were

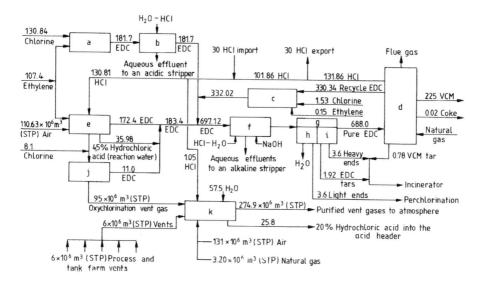

Figure 69. VCM production according to the HCl-balanced old process
a) Direct chlorination; b) Acidic wash; c) Recycle EDC chlorination and dechlorination; d) VCM cracking; e) Oxychlorination; f) Acidic and alkaline wash; g) EDC distillation; h) Light ends and H_2O removal; i) Heavy ends and tar removal; j) Chlorination of vent gas from oxychlorination; k) Vent gas incinerator with HCl recovery (Figures are given in 10^3 t/a unless otherwise specified)

incinerated internally or partly externally (depending on the uptake capacity of the rotary kiln incinerator) with heat recovery but without utilization of their considerable chlorine content. This resulted in a high salt load in the flue gas scrubber effluent. The light ends, ca. 36% of the total residue amount, were used as raw material for the production of tetrachloroethylene by a perchlorination process. Because this perchlorination process also produced hexachlorobenzene- (HCB) and hexachlorobutadiene- (HCBD) containing wastes— 3600 t of light ends form 345 t of those solid wastes— which had to be disposed at great expense on an underground deposition site, this way of utilizing light ends finally was only a transposed residue recycling without a really closed loop.

The large amounts of toxic vent gases [containing, besides partly cancerogenic chlorohydrocarbons (CHCs), compounds such as carbon monoxide, chlorine, HCl, and hydrocarbons] are produced predominantly during oxychlorination with air. Further emissions are due to various process vent gases and tank breathing. Despite the large gas volume, these vent gases have been treated since 1977 by thermal incineration in an integrated HCl recovery plant (first plant in Europe, Fig. 70) because catalytic incineration, which consumes much less energy, was not practicable due to the high chlorine content of these vent gases (chlorine acts as a strong catalyst poison). Based on the low calorific value of these vents, considerable amounts of natural gas had to be used as auxiliary fuel to guarantee complete carbon combustion at 950 °C and

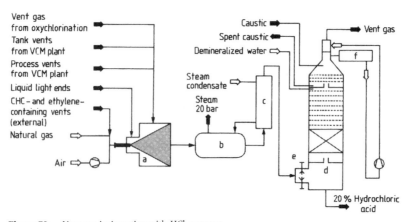

Figure 70. Vent gas incineration with HCl recovery
a) Combustion chamber; b) Waste-heat boiler; c) Boiler feedwater preheater; d) Wash column; e) Quenching vessel; f) Demister

0.6 – 0.8 s residence time. Most of the waste heat of the flue gas was utilized for the generation of steam at 20 bar by a waste-heat boiler. The mostly organically bonded chlorine in the vents was recovered as 20 % hydrochloric acid in the two-stage flue gas scrubber. To satisfy the site demand of 20 % hydrochloric acid, additional crack HCl had to be introduced into the vent gas incinerator.

Wastewaters produced in the different process stages are (1) reaction water of the oxychlorination process; (2) $FeCl_3$-containing wastewater from the acidic wash decanter of the direct chlorination; (3) slightly alkaline wastewater from azeotropic drying of wet crude EDC in the light ends column; (4) sodium-hypochlorite-containing spent caustic from the flue gas scrubber of the incineration plant; (5) strongly alkaline wastewater from the VCM wash decanter and VCM drying traps; and (6) wastewater from the acidic and alkaline wash decanter of the oxychlorination unit with high COD. Additional surficial wastewater is collected in plant and tank farm pits. This wastewater is more or less strongly contaminated with CHCs. An exception is the spent caustic from the incineration plant. The chlorinated hydrocarbons dissolved in wastewater were recovered by steam stripping and recycled back to the process. Three steam strippers were in operation: one for the surficial wastewater collected in several pits; a second stripper for $FeCl_3$-containing wastewater from the acidic wash decanter of direct chlorination (which has a low COD); and a third stripper (oxy stripper) for acidic reaction water from oxychlorination (which has a high COD), for the COD-rich wastewater from the acidic or alkaline wash decanter of the oxychlorination, for wastewater from the azetropic drying of the wet crude EDC, and for wastewater from the VCM wash decanter and drying traps. The effluents of strippers 1 and 2 were led into a chemical – mechanical sewage treatment plant ($FeCl_3$ is a desired flocculant in the neutral or slightly alkaline range); the alkaline salt-containing oxy stripper effluent was led into a one-stage biological sewage treatment plant. The effluent from the oxy stripper was still contaminated with a relatively high COD and adsorbable organic halogen (AOX) load.

Besides this high energy-consuming wastewater pretreatment, the combined VCM production process has a high steam consumption in the various processing steps that are necessary for thermal separation of the diverse intermediate material mixtures. Despite steam credit for the integrated heat recovery in the central vent gas incineration plant and for recovery of reaction heat from the exothermic oxychlorination reaction (generation of steam at 20 bar) and the exothermic chlorination of oxychlorination vent gas (generation of steam at 1.5 bar), the specific steam consumption amounts to 1.74 t per tonne of VCM with 55 % conversion rate at EDC cracking.

The consumption of electrical energy for pumps, for the refrigeration unit, for the air compressor, and for the combustion air and flue gas suction fan of the vent gas incineration plant was 136 kW $\cdot$ h per tonne of VCM.

Large amounts of natural gas are necessary for firing of the EDC cracking furnace, and for the generation of radiant heat as heat flux for the endothermic cracking process, and as auxiliary fuel for the vent gas incineration plant. Thereby, most of the sensible heat of the superheated reaction product leaving the cracking furnace and of the waste heat of the flue gases used for firing the cracking furnace was lost to the environment, resulting in a specific natural gas consumption of 954 kW $\cdot$ h per tonne of VCM.

Despite the close integration of the VCM process with chlorine-producing and chlorine- or HCl-consuming plants, the old VCM production process was not quite satisfactory in economical or ecological terms.

New Process for Integrated VCM Production. The process changeover aimed at the following objects:

1) Increase of yields, i.e., minimization of formation of by-products by more selective reactions
2) Utilization of all CHC-containing residues produced in the integrated process with chlorine recycling and heat recovery in a totally closed material recycling loop
3) Reduction of wastewater flow with simultaneous minimization of COD, AOX, and salt load
4) Energy savings in the distillation processing of the particular material streams, and in the EDC cracking unit
5) Energy savings in the vent gas incineration unit by import of external CHC-containing vent gases with high calorific value and by improving the thermal efficiency of the vent gas incineration unit.

Figure 71 depicts the overall process flow and mass balance of the new process built at the end of 1991. The new process became "HCl unbalanced" by crack HCl export to silicon chemistry after special purification [171] and simultaneous shutdown of the perchlorination process for economical and ecological reasons. All residues containing chloroorganic compounds are thermally treated under chlorine and heat recovery in the new HCl generation plant (Fig. 72) [172].

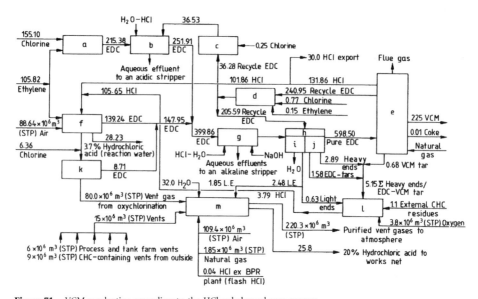

Figure 71. VCM production according to the HCl-unbalanced new process
a) Direct chlorination; b) Acidic wash; c) Benzene removal; d) Recycle EDC chlorinaton and dechlorination; e) VCM cracking; f) Oxychlorination; g) Acidic and alkaline wash; h) EDC distillation; i) Light ends and H_2O removal; j) Heavy ends and tar removal; k) Chlorination of vent gas from oxychlorination; l) By-product recycling (BPR) plant (oxygen incinerator); m) Vent gas incinerator with HCl recovery (Figures are given in 10^3 t/a unless otherwise specified)

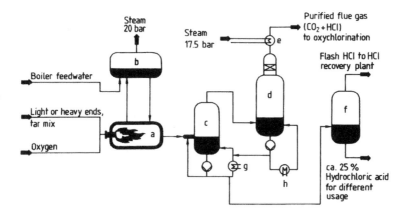

Figure 72. HCl-generating plant
a) Reactor (combustion chamber); b) Steam drum; c) Quenching vessel; d) Washer; e) Heater; f) Flash drum; g) Cooler; h) Cooler (brine at −15 °C)

The process is based on the following chemical reaction:

$$C_xH_yO_zCl + \left(x + \frac{y-1}{4} - \frac{z}{2}\right)O_2$$
$$\rightarrow x\,CO_2 + HCl + \left(\frac{y-1}{2}\right)H_2O$$
$$\Delta H^R_{25°C} = -12\,560 \text{ to } -14\,650 \text{ J/kg}$$

Because of an H:Cl atomic ratio of >1.3:1 in the residues to be incinerated, the reaction is autothermal and needs no auxiliary fuel.

With a 4–6% oxygen excess related to stoichiometry—required for optimum carbon combustion, i.e., to avoid CO formation—the temperture dependent Deacon equilibrium reaction is shifted to the right:

$$4\,HCl + O_2 \rightleftharpoons 2\,Cl_2 + 2\,H_2O$$

The following reaction products are formed:
1) Flue gas with the following composition is fed to the oxychlorination unit without inert gas separation:

CO_2	56.8 vol%
Cl_2	1.1 vol%
HCl	32.6 vol%
O_2	3.2 vol%
N_2	6.3 vol%

2) 25% hydrochloric acid for general usage
3) Flash HCl which is sent to the HCl recovery plant in the gaseous state

The chlorine recovery rate as gaseous HCl is ca. 90%. The steam production is ca. 4.4 t per tonne of residue, which amounts to ca. 85% heat recovery related to the calorific value of the residues.

The CHC residue mix is converted to gaseous HCl by reaction with oxygen at a pressure of 8 bar gauge and at an extremely high temperature (up to 2500 °C in the flame). The HCl formed is recycled to the oxychlorination unit without expensive compression or inert gas separation steps [172]–[174]. Use of pure oxygen instead of air has the advantages of an extremely high flame temperature and very long residence times in the combustion chamber. This guarantees (1) quantitative conversion of CHC to CO_2, H_2O, and HCl (in addition to traces of chlorine according to the Deacon equilibrium reaction), and (2) complete destruction of polychlorinated biphenyls (PCB) and polychlorinated dibenzodioxins and dibenzofurans (PCDD, PCDF) being present in trace amounts in the heavy ends and EDC–VCM tars [175] and also preventing their de novo synthesis (destruction rate for PCB: 99.99999%, for PCDD/PCDF: 99.992%). By operating this HCl generation plant, the material recycling loop is totally closed. To save auxiliary fuel (natural gas) and satisfy the internal site demand for 20% hydrochloric acid, part of the liquid light ends and fuel-rich CHC-containing vent gas from other plants are incinerated in addition to the VCM vent gases in the central HCl recovery plant with heat recovery. By installation of a boiler feedwater preheater [176], the thermal efficiency of the vent gas incinerator was strongly enhanced and the water vapor content in the purified flue gas drastically reduced. Even though liquid light ends were incinerated in addition, the PCDD–PCDF content in the purified flue gas is less than 0.1 ng/m^3 (STP) calculated as international toxicity equivalents (ITEQ). Besides, the so-called fugitive emissions were extensively reduced by technical measures such as canned pumps, grooved gaskets, emission-free sampling devices, and sucking off open drains and wastewater collection pits in the direction of the vent gas incinerator. Thus, the total output of toxic compounds into air could be reduced by 90 % in combination with technical measures for increasing the on-stream factor of the HCl recovery plant (e.g., by improving burner control and selecting more resistant construction material). Improvements in the EDC washing unit [177] led to a drastic reduction of the COD load of the wastewater. Process improvements in the oxy stripper led to steam savings and a nearly quantitative hydrolysis of chlorinated ethanol to glycol and glycolaldehydes, followed by essential reduction of wastewater AOX load. The overall stripping efficiency was increased drastically. Currently, surficial wastewater is collected in a 200-m^3 basin and purified in an adsorber (Fig. 73). The new process saves energy (steam consumption only in the desorption step) and is very efficient in terms of AOX reduction and elimination [178]. All these improvements resulted in a higher overall purification efficiency with a reduction of the total wastewater flow of about 25 % (see below):

Old process
Alkaline stripper for COD-containing wastewater
Acidic stripper for inorganic wastewater (FeCl$_3$ containing) with low COD
Neutral stripper for surficial wastewater

Wastewater flow:	235 577 t/a
CHC recovery:	2 538 t/a
CHC composition:	46.7 wt % 1,2-C$_2$H$_4$Cl$_2$
	49.2 wt % CHCl$_3$
	2.0 wt % C$_2$H$_5$Cl

Remainder: other chlorinated C$_1$–C$_4$ hydrocarbons
Purification efficiency: 99.75 %

New Process
Improved alkaline stripper for COD-containing wastewater
Improved acidic stripper for inorganic wastewater (FeCl$_3$ containing) with low COD
Adsorptive purification of surficial wastewater at pH ≈ 1

Wastewater flow:	178 780 t/a
CHC recovery:	1 156 t/a
CHC composition:	75.2 wt % 1,2-C$_2$H$_4$Cl$_2$
	22.0 wt % CHCl$_3$
	1.6 wt % C$_2$H$_5$Cl

Remainder: other chlorinated C$_1$–C$_4$ hydrocarbons
Purification efficiency: 99.95 %

Simultaneously, the biological sewage treatment plant was extended by a second stage that additionally treats municipal wastewaters (improving the AOX elimination rate) and by a biosludge (fluidized-bed) incinerator with flue gas purification. The CHC of the return cooling water and steam condensate of the VCM plant is measured by on-line analyzers (semiconductor sensor, FID) to recognize small leakages very quickly. In connection with operating instructions for directives in case of a leak, CHC input into the main outfall ditch because of leakages is drastically decreased. All these measures contribute to keeping the AOX content in the effluent of the biological sewage treatment plant distinctly below the existing regulatory emission levels (max. 20 g AOX per tonne of pure EDC or max. 1 mg AOX/L, or max. 5 g EDC per tonne of pure EDC or 2.5 mg EDC/L, both as monthly averages).

Production-integrated improvements have been achieved by long-term research in the field of chemical catalysis and radical chain pyrolysis. The improvements are more selective catalysts for oxychlorination with increased activity [179] (fewer by-products, higher HCl conversion, lower degree of ethylene oxidation rate), an improved catalyst for chlorination of the oxychlorination vent gases [180] (nearly quantitative ethylene conversion by increased activity and, because of increased selectivity, much lower formation of chlorinated ethanols that interfere with the VCM process); and a totally changed heat flux system for EDC pyrolysis [181] which, by using the so-called reverse heat flux technique, reduces the amount of by-products and coke produced even at cracking depths around 60%. By chlorination of a small portion of recycle EDC a benzene concentration of <500 ppm in the feed EDC could be achieved [182]. This influences the cracking kinetics favorably and less coke and VCM tar were produced. Chloroprene removal from total recycle EDC was distinctively improved [183]. Further ressource saving and by-product lowering effects were obtained by improving acetylene removal from cracking HCl by selective hydrogenation [184] and by amending the reaction parameters of direct chlorination [185].

With the pinch technology, important energy savings were obtained. At first, all material streams had to be summed up. Based on this, hot process streams with a technically utilizable temperature level were coupled with heat-absorbing process steps [181], [186]. Also the steam consumption of EDC distillation and EDC vaporization into the cracking furnace could be reduced further [177], [181], [183]. In combination with an open heat pump for reboiling the especially energy-consuming heavy ends column (so-called vapor thermocompression) [187], [188], the specific steam consumption of the VCM process has been lowered to 0.25 t per tonne of VCM including steam production in the oxychlorination unit and in the HCl recovery and HCl generation plant. Further steam could be saved as a side effect of the thermocompression in the VCM column, in the liquid feed EDC preheat before entering the cracking furnace, and in the EDC vaporization into the cracking furnace. The increased electrical power consumption for vapor recompression (ca. 45 kW · h per tonne of VCM) is more than compensated by the overall energy savings. These overall energy savings reduce the consumption of natural gas by a total amount of $24.5 \times 10^6 m^3$ (STP)/a, resulting in less emission of CO_2 (51 940 t/a) and $(NO)_x$ (21.9 t/a).

Ecological and Economical Evaluation of the New Process. By the economically and ecologically significant utilization of residues from the integrated VCM production process, an amount of chlorine is saved that corresponds to the flow of recycled HCl. In addition, HCl or salt-containing wastewaters are avoided. Due to the combination of

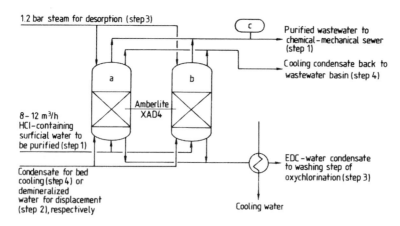

Figure 73. Flow sheet of an adsorber plant (specific steam consumption: 42 kg per cubic meter of wastewater, 70 % less than a steam stripper with feed-bottom heat exchanger)
a) Working adsorber, 2 m³; b) Standby adsorber, 2 m³; c) FID analyzer for breakthrough determination

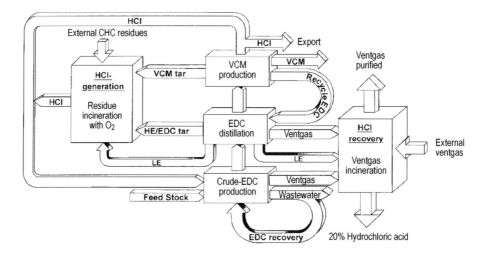

Figure 74. Material recycling loops and residues and ventgas utilization in VCM production
HE = Heavy Ends; LE = Heavy Ends

the HCl recovery and generation plant and the AOX pretreatment of wastewater with recycling of CHC removed from wastewater, the new process for manufacturing EDC and VCM is nearly emission and residue free. The HCl generation plant makes the VCM process totally self-sufficient in terms of residue utilization. In combination with measures for process optimization that save ressources and energy, the specific input data and the environmental pollution balance could be demonstrably improved in relation to the old process (Table 4). The potential for the formation of chlorinated dibenzodioxins and -furans has been reduced. Remaining amounts are destroyed almost quantitatively. A shift of dioxin emissions to residues and wastewater is therefore

Table 4. Comparison of specific input and output data in VCM production (Target products are each 225 000 t/a VCM, 30 000 t/a HCl for export, and 25 800 t/a 20% hydrochloric acid for total site demand)

Substance or energy	Old process	New process	Change, %
Input			
Ethylene, kg per tonne VCM	478	471	−1.5
Chlorine[a], kg per tonne VCM	628.8	592.4	−5.8
Electrical power, kW · h per tonne VCM	136	188	+38.2
Steam, t per tonne VCM	1.74	0.25	−85.6
Natural gas, kW · h per tonne VCM	954	835	−12.5
20% Hydrochloric acid, kg per tonne VCM	6	3	−50.0
Caustic[b], kg per tonne VCM	40	15	−62.5
Process water[c], m^3 per tonne VCM	0.86	0.56	−34.9
Output			
Coke, g per tonne VCM	90	45	−50.0
Wastewater, m^3 per tonne VCM	1.05	0.77	−26.7
COD, kg per tonne VCM	15	1.3	−91.3
AOX, kg per tonne VCM	0.35	0.14	−60.0
NaCl[b], kg per tonne VCM	50	15	−70.0
Vent gas purified, m^3 (STP) per tonne VCM	1220	979	−19.8
Residue (heavy ends, tar), kg per tonne VCM	28	0	−100.0
CO_2[d], kg per tonne VCM	776.7	588.7	−24.2
$(NO)_x$ as NO_2[d], g per tonne VCM	585.9	437.9	−25.3
$\sum$ Toxic compounds (air), kg per tonne VCM	4.9	0.5	−89.8

[a] Including HCl export–import and production of 20% hydrochloric acid.
[b] Neutralization of oxychlorination reaction water, wastewater of EDC acidic wash, surficial wastewater, or acidic wastewater from flue gas scrubber of heavy ends–tar incineration.
[c] Demineralized water, however, without boiler feedwater for internal and external steam production.
[d] Integrated vent gas and liquid residue incineration, external residue incineration (old processes), cracking furnace firing, and electrical power and steam production in a thermal power station required for VCM production.

avoided. These successfully operating production-integrated improvements for environmental protection are fundamentally based on the combined material balance with closed-loop material recycling (Fig. 74). High investment costs were necessary for the chemical engineering of the new integrated process, including new equipment, and for proper design of the operation and reaction control parameters supported by a modern computerized process control system with video terminals. Extensive research and development work over years required additional capital costs for thorough investigation of each process step.

3.2.5. Examples from Hüls

3.2.5.1. Integrated Environmental Protection in Cumene Production

Introduction. Hüls has produced cumene since 1954 at its site in Marl. The current production capacity is 150 000 t/a. Cumene is used mainly in the production of acetone and phenol, and also in the production of cumene sulfonate, which is a raw material for detergents. It is produced by Friedel–Crafts alkylation of benzene by propene, catalyzed by aluminum chloride, which does not give cumene exclusively, but also produces di- and tiisopropylbenzene.

$$C_3H_6 + C_6H_6 \xrightarrow{AlCl_3} C_6H_5\text{-}C_3H_7$$

$$2\,C_3H_6 + C_6H_6 \xrightarrow{AlCl_3} C_6H_4(C_3H_7)_2 \quad (meta \text{ and } para \text{ isomers})$$

$$3\,C_3H_6 + C_6H_6 \xrightarrow{AlCl_3} C_6H_3(C_3H_7)_3$$

Other side reactions also occur to a small extent. Whereas the polyalkylated isopropylbenzenes can be converted to the target product cumene by using aluminum chloride as the catalyst, other by-products, which include a high proportion of ethylbenzene, must be removed from the reaction mixture.

$$C_6H_3(C_3H_7)_3 + C_6H_6 \xrightarrow{AlCl_3} C_6H_4(C_3H_7)_2 + C_6H_5\text{-}C_3H_7$$

$$C_6H_4(C_3H_7)_2 + C_6H_6 \xrightarrow{AlCl_3} 2\,C_6H_5\text{-}C_3H_7$$

Conventional Method of Producing Cumene (Fig. 75). In the past, the Friedel–Crafts alkylation of benzene was carried out together with the conversion of the polyalkylated isopropylbenzenes in a single reactor (a). The catalyst was produced directly in the reaction mixture from aluminum chippings and HCl. It formed a second separate organic phase in the reactor. The water present in the propene was simply removed mechanically, whereas the benzene used was dried over sodium hydroxide.

After removal from the reactor, the catalyst was decomposed by addition of water, and the liberated aluminum chloride was extracted (b) into the aqueous phase. However, the amount of water required was determined not by the extraction parameters but by the

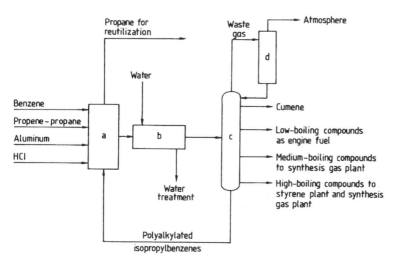

Figure 75. Cumene production—old process: a) Reactor; b) Extraction; c) Distillation; d) Purification

decrease in temperature required for the water washing. Thus, a high excess of cold water was added to the hot reaction mixture to obtain the desired mixing temperature.

Subsequent distillative processing (c) led to the following

1) The target product cumene
2) Polyalkylated isopropylbenzenes, which were returned to the reactor
3) Low-boiling products, which were used as motor gasoline
4) Medium-boiling products, which were processed within the Hüls organization to produce synthesis gas
5) High-boiling products, which were used first as solvents in the production of styrene and then as raw materials for the production of synthesis gas

New Process for Cumene Production [189] (Fig. 76). The new process for the production of cumene was introduced between 1986 and 1988. Its most important features were an increase in the selectivity of cumene formation and suppression of undesired side reactions that do not lead to monoalkylated or polyalkylated isopropylbenzenes. This was achieved by the following modifications to the process:

1) Catalyst production in a separate reactor (a)
2) Avoidance of a second organic catalyst phase in the alkylation reactors (b)
3) Separation of the alkylation of benzene by propene from the reaction of the recycled polyalkylated isopropylbenzene with benzene
4) Complete dehydration of the starting materials benzene and propene by distillation

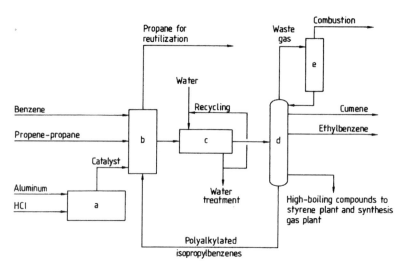

Figure 76. Cumene production — new process
a) Catalyst production; b) Reactors (several stages); c) Extraction; d) Distillation; e) Purification

An improvement in the distillative processing of the reaction mixture enabled ethylbenzene to be isolated in a pure state and used as starting material in styrene production.

Because of the total dehydration of the starting materials, catalyst consumption and the associated contamination of wastewater by inorganic substances were reduced. The amount of wastewater could be lowered by separating the cooling of the reaction mixture from aqueous extraction of the catalyst. Since the low temperature necessary for extraction could now be adjusted by indirect cooling, the amount of wastewater produced is determined only by the operating parameters for the extraction. The water produced in the extraction can be recycled many times.

Due to the introduction of heat recovery in the distillation process, the energy consumption of the entire process was decreased. Whereas formerly the heat of condensation was removed without utilization, it is now used to produce process steam and hot water, which are fed to the supply networks of Hüls at Marl.

Results. A comparison of the old cumene process (1980) with the new (1990) shows that the amount of wastewater is now 5 % of the former amount, the organics content of the wastewater after passing through the water treatment plant is 1 % of its former value, and the amount of residues has been reduced to 50 % of the earlier figure. Energy consumption, despite additional requirements for distillative processing of the starting materials, has been reduced by 60 %. The gaseous emissions of organic substances, which were already very low, have been prevented completely by passing the waste gases through an existing incinerator.

3.2.5.2. Production of Acetylene by the Hüls Plasma Arc Process

Production and Use of Acetylene. As a result of its strongly unsaturated character, acetylene reacts readily with many compounds. During the early development of industrial organic chemistry, it was one of the most important starting materials for a large number of products. Since the 1940s (United States) and the 1950s (Europe), it has been largely replaced by ethylene and other olefins. Bulk products still produced from acetylene include butanediol and vinyl compounds (vinyl chloride, vinyl ethers, and vinyl esters).

An important feature of acetylene production is the high energy requirement due to the formation of a triple bond. In the Hüls plasma arc process, the energy is supplied by an electric arc. Hydrocarbons are fed to the electric arc reactor in gaseous form and are pyrolyzed at high temperature (>2000 K). At these temperatures, acetylene is formed in preference to other hydrocarbon compounds. To suppress further reaction to carbon and hydrogen, the hot crack gas must be cooled rapidly (a few microseconds) to low temperatures.

In the electric arc reactor, which has a power consumption of 8–10 MW, ca. 1 t/h acetylene is produced from ca. 2 t/h hydrocarbons. The composition of a typical reaction gas mixture is given below.

C_2H_2	16 vol %
H_2	60 vol %
CH_4	10 vol %
C_2H_4	8 vol %
C_3 compounds	2 vol %
C_4 compounds	1.5 vol %
Higher alkynes	1.5 vol %
Residues	1 vol %
Carbon black	0.4 kg per kilogram C_2H_2

The hot reaction gas with this composition is processed in the following main steps:

1) Cooling
2) Removal of carbon black, higher-boiling hydrocarbons (aromatics), and polyunsaturated compounds
3) Purification of acetylene and by-products such as hydrogen and ethylene

Hüls has produced acetylene since 1939 by electric arc technology. Until acetylene was superseded by ethylene as starting material for poly(vinyl chloride) manufacture in 1992, the production capacity was 12×10^4 t/a using 18 electric arc groups. After this changeover, the supply of the remaining consumers has been satisfied by using a new method of processing the crack gas that was developed jointly with the company Linde. This method meets the stricter present-day ecological and economic requirements. This was achieved mainly by techniques of production-integrated environmental protection.

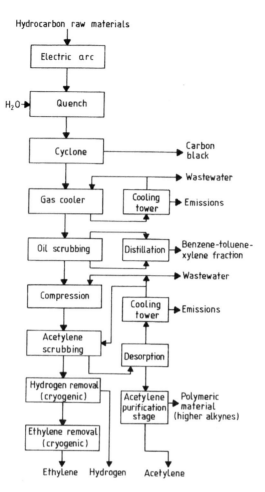

Figure 77. Flow diagram of the old Hüls acetylene process

Old Production Process. The relevant steps of the old process are shown in the flow diagram in Figure 77. Hot crack gas from the electric arc furnace was quenched to ca. 200 °C with water to enable most of the carbon black to be removed in cyclones. The quench water condensed in a further scrubbing step with water at ca. 50 °C, and the remaining carbon black was removed at the same time. After mechanical removal of carbon black the heated wash water was cooled by direct contact with air in a cooling tower and recycled to the cooler. Excess water was fed to the wastewater treatment plant. Both carbon black fractions recovered were used as fuel.

The high-boiling, mainly aromatic compounds were washed out with special oil fractions and then recovered by distillation.

The crude gas so obtained was compressed; acetylene was absorbed at increased pressure in water and then desorbed as crude acetylene for further purification by decompression. The wash water was fed into the recycling system, some of it being cooled by direct contact with air. The remaining hydrocarbons in the crude gas were recovered by cryogenic processes involving gas liquefaction.

The polyunsaturated and hence energy-rich compounds recovered in the acetylene purification stage were thermally treated, and the polymerizate so obtained could be used as fuel without problems.

The main reasons for the high emissions and high wastewater contamination were the large, partly open water recycling systems. The emissions consisted of hydrocarbons and HCN formed in the electric arc reactor from the reaction of the nitrogen present in the starting materials.

Of the electrical energy supplied, ca. 50 % was utilized for chemical reactions. The rest (thermal energy) was not used.

New Production Process. The flow diagram of the new plant is given in Figure 78. The main difference between the new and the old processes is that the hot reaction gas from the electric arc furnace is quenched with water only to the temperature level at which it is cool enough for no further reaction of acetylene to carbon black and hydrogen to occur on subsequent cooling in a heat exchanger.

The thermal energy of the hot reaction gases is used to produce hot steam in a heat exchanger of special design and operational features that prevent the formation of deposits by the hot carbon black present in the gas.

The carbon black is recovered mainly in cyclone separators, and the gas is further cooled by washing with oil, which also removes the high-boiling aromatics. Heavy fuel oil is used as the washing medium, and this, together with the remaining carbon black, is used to produce synthesis gas. The "dry carbon black" from the cyclone separators is used as fuel in the power station.

The reaction gas is cooled to ambient temperature in a water spray column, and the quench water that condenses is fed back in a closed cycle to the electric arc reactor. The gas is then compressed and subjected to a sequence of scrubbing operations with selective solvents [i.e., methanol, octane, and *N*-methyl-2-pyrrolidone (NMP) in that order]. The solvents are regenerated by low-pressure stripping with gas fractions from the production process. This removes mainly the polyunsaturated hydrocarbons.

The acetylene-free gas stream from the NMP wash is first passed through a cryogenic separating stage (cold box) which removes most of the remaining hydrocarbons. These are used first as stripping gas in the regeneration stages of the washing processes described above and are then recycled to the electric arc process. The raw hydrogen stream from the cold box is processed in the pressure swing adaptation (PSA) plant to pure hydrogen and waste gas for heating purposes.

The formation of HCN is almost completely prevented by using virtually nitrogen-free hydrocarbon raw materials. Small residual amounts of HCN are effectively removed by scrubbing with NaOH.

Improvements Resulting from the New Process. The new process is characterized by a high degree of heat recovery and the use of closed cycles. This gives the following improvements:

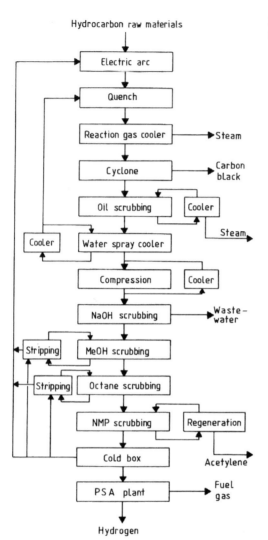

Figure 78. Flow diagram of the new Hüls acetylene process

1) Increase in the thermal efficiency from ca. 50 % to the present figure of 75 %
2) Drastic reduction in the amount of wastewater to be treated in the wastewater treatment plant
3) Prevention of emissions of hydrocarbons and HCN
4) Utilization of the residual carbon black and polyunsaturated hydrocarbons that are recovered in the scrubbing processes

Considerable economic benefits have been obtained:

1) Lowering of energy consumption and wastewater costs
2) Lower personnel requirements
3) Improvement in product purity

3.2.6. Low-Residue Process for Titanium Dioxide Production (Example from Kronos International)

General. Titanium dioxide pigment is an important industrial intermediate for multiple utilization in various areas (paints and coatings, printing inks, plastics, man-made fibers, paper, etc.) [190]. The raw materials for pigment production contain titanium dioxide associated with oxides of other metals, mainly iron. The only way to isolate titanium dioxide from this oxide mixture in pigment-grade purity is by chemical processing.

The first commercial process, the sulfate process, uses concentrated sulfuric acid as reagent. After use, the acid cannot readily be recycled because of its content of metals and water taken up during the process. Therefore, this acid (so-called spent acid) was typically discarded as waste. Because of the large acid quantities involved, the (conventional) sulfate process became increasingly obsolete and had to be improved to a modern version [191].

An alternative to the sulfate process is the chloride process for titanium dioxide pigment manufacture, using chlorine as reagent [192], [193]. Since regeneration and recycling of chlorine after use have always been an integral part of this process, it never was impaired by a problem such as the disposal of spent acid from the sulfate process. Thus, the chloride process has become more and more successful and has already gained equal or even greater importance than the sulfate process, although it was developed considerably later.

Conventional (Sulfate) Process. Raw material is either an ilmenite ($FeTiO_3$) concentrate from beneficiation of naturally occurring ilmenite or a slag obtained from ilmenite by electrosmelting and tapping off the major part of the iron as pig iron for further use. The titaniferous material is digested with concentrated sulfuric acid, leading to a solution of sulfates of most of the metals contained in the ore, e.g.:

$FeTiO_3 + 2\,H_2SO_4 \rightarrow TiOSO_4 + FeSO_4 + 2\,H_2O$

The insoluble solids (unreacted ore, gangue) are separated, and part of the iron(II) sulfate is crystallized and recovered (this part of the process generally being omitted when starting from slag). Subsequently, titanium is selectively precipitated by hydrolysis

$TiOSO_4 + 2\,H_2O \rightarrow TiO(OH)_2 + H_2SO_4$

Subsequent filtration yields the spent acid filtrate and a cake of aqueous titanium oxide. Thorough washing of the cake, calcination, grinding, surface treatment in suspension, filtration, drying, and micronizing finally lead to the titanium dioxide pigment product. These process steps result in some slightly acidic effluents.

The spent acid obtained by the filtration step after hydrolysis is diluted sulfuric acid, 18–23 wt% H_2SO_4, containing dissolved sulfates of iron and other metals from the raw material. In the conventional process, ca. 6–9 t of spent acid is generated per tonne of pigment, depending on the raw material used. The spent acid was discharged into rivers, coastal waters, or the high seas [194]. Formerly, the recovered iron(II) sulfate (when using ilmenite as raw material) was redissolved in the spent acid and discarded as well.

Modern Sulfate Process. The sulfate process was thoroughly optimized to reduce the quantity of spent acid and increase its H_2SO_4 concentration. Nevertheless, a substantial amount of spent acid remained; its low H_2SO_4 concentration and its metal sulfate content prevented direct recycling to digestion. Further improvement of the conventional sulfate process by alterations proved impossible, and attempts to commercialize a novel sulfate process with direct recycling of all the spent acid failed. Therefore, further processing of the spent acid was mandatory. In some countries, the spent acid is concentrated somewhat and then used for fertilizer production. In other countries, especially those that have few natural gypsum deposits, the spent acid is reacted with calcium compounds to produce gypsum for use (e.g., in plasterboard manufacture).

Another solution to the spent acid problem is improving its quality by concentration and purification to the extent that it can be completely recycled to the digestion step in pigment production. This alternative is adopted, for example, by all producers in Finland, Germany, and Spain. Although individual plants may differ in some details, the multiple-step reprocessing and recycling of spent acid generally consist of [192], [195]–[197]:

1) Concentrating the spent acid up to 60–75 wt% H_2SO_4 where the dissolved metal sulfates have a minimum solubility and are mostly crystallized.
2) Separation of the solid metal sulfates (so-called filter salt) from the acid by filtration. Sometimes, the acid filtrate can be recycled directly to digestion; otherwise, further concentration up to 80–90% H_2SO_4 is required before recycling.

The separated metal sulfates can be processed further by thermal decomposition to generate metal oxides (mainly iron oxide) and sulfur-dioxide-containing gas, which is then processed to virgin sulfuric acid as in conventional pyrite roasting–sulfuric acid plants.

When ilmenite is used as raw material for pigment production, the spent acid contains large amounts of dissolved iron(II) sulfate. In this case, spent acid treatment starts with process steps permitting the recovery of a major part of the iron(II) sulfate before further concentration. This iron(II) sulfate or copperas, $FeSO_4 \cdot 7\,H_2O$, is marketed jointly with the copperas recovered during titanium dioxide production. In Germany, for example, none of the copperas produced is discarded any longer; rather, it is utilized, as such or after further processing, for multiple environmental purposes (water treatment, effluent or sludge treatment in sewage plants).

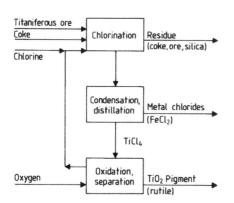

Figure 79. Flow diagram of the chloride process for production of titanium dioxide pigment

Chloride Process. The chloride process (see Fig. 79) was developed and commercialized as an alternative to the sulfate process. The processing of titaniferous raw materials (ilmenite, slag, or rutile) occurs via the metal chlorides, by using chlorine as reactant and carbon as a reducing reagent, e.g.:

$$TiO_2 + FeO + 3\,Cl_2 + 2\,C \rightarrow TiCl_4 + FeCl_2 + CO + CO_2$$

This exothermic reaction takes place at ca. 1000 °C. The reaction products are gases, and after fractional condensation, solids separation, and distillation, sufficiently pure, liquid $TiCl_4$ is produced. The $TiCl_4$ is then reacted with pure oxygen to generate titanium dioxide pigment:

$$TiCl_4 + O_2 \rightarrow TiO_2 + 2\,Cl_2$$

Chlorine is separated from the solid pigment and directly recycled to the first step, chlorination. The pigment may be surface treated in suspension and further processed in the same way as sulfate pigment.

Ecological Aspects of Modern Production Processes. *Sulfate Process.* The conventional sulfate process is characterized by a linear flow of sulfuric acid through the process. Some H_2SO_4 ends up in the copperas by-product, but the main part remains separate from the pigment end product, as a used reagent with deteriorated quality in terms of concentration and purity. Formerly, this large stream was discarded as waste (see Fig. 80). Now, the spent acid is recovered, and complex acid concentration and filter salt treatment plants are added to recycle the acid entirely. Hence, the modern sulfate process plant for titanium dioxide pigment manufacture is characterized by a closed sulfuric acid cycle that completely withholds spent acid from the environment (see Fig. 81).

This production-integrated recovery could not be achieved by process redesign; new, separate process units had to be added to the essentially unchanged conventional manufacturing plant. Thus, not only did closing of the acid cycle require substantial investment, but energy requirements increased considerably as well. A substantial amount of fossil fuel is required as such—for thermal decomposition of the filter

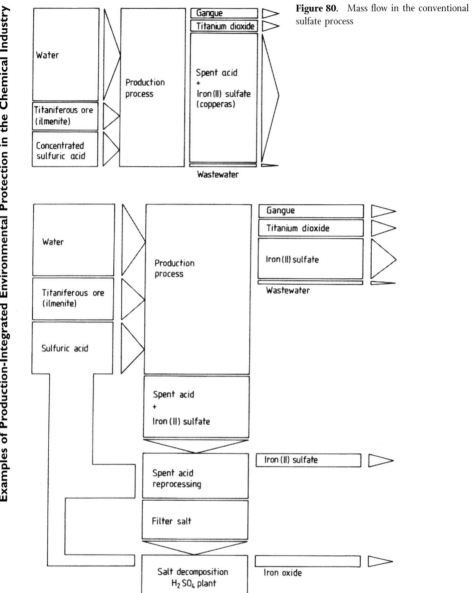

Figure 80. Mass flow in the conventional sulfate process

Figure 81. Mass flow in the modern sulfate process

salt—or indirectly in the form of valuable steam for water evaporation in acid concentration. Parts of these additional energy requirements cannot be recovered but are lost to the environment via cooling towers and off-gases.

Chloride Process. In contrast to the sulfate process, the chloride process has the great benefit that the reagent chlorine is recycled directly without major treatment, this recycling being an integral part and an inherent characteristic of the process right from the beginning. In past years, the chloride process has increasingly been used in construction of new titanium dioxide pigment factories or even for replacing obsolete sulfateprocess-based plants to cope with the spent acid problem.

Process Engineering Equipment Aspects. Spent acid treatment and concentration as part of the modern sulfate process were possible only because of the availability of recently improved construction materials and equipment items permitting the safe handling of this highly corrosive substance.

Chloride process technology, plant design, and equipment are entirely different from sulfate process technology. Its commercialization required considerable development and engineering efforts. Safe operation is possible because of high-technology equipment and sophisticated control technology.

3.2.7. Reduction of Waste Production and Energy Consumption in the Production of Fatty Acid Methyl Esters (Example from Henkel)

Fats and Oils: The Raw Materials of Oleochemistry. Fats and oils are triglycerides (i.e., fatty acid esters of glycerol). They are the starting materials for the production of fatty acid methyl esters, which are important intermediates in the production of fatty alcohols and surfactants [198] by the oleochemical route, which has great ecological benefits [199]. The fatty acid methyl esters are produced either by the esterification of fatty acids after hydrolysis of the triglycerides or by direct transesterification with methanol. The overall transesterification reaction is as follows:

$$\begin{array}{c} \text{CH}_2-\text{O}-\overset{\overset{\text{O}}{\|}}{\text{C}}-\text{R} \\ | \\ \text{CH}-\text{O}-\overset{\overset{\text{O}}{\|}}{\text{C}}-\text{R} \;+\; 3\,\text{CH}_3\text{OH} \;\overset{\text{Alkali}}{\rightleftharpoons}\; 3\,\text{CH}_3-\text{O}-\overset{\overset{\text{O}}{\|}}{\text{C}}-\text{R} \;+\; \begin{array}{l} \text{CH}_2-\text{OH} \\ \text{CH}-\text{OH} \\ \text{CH}_2-\text{OH} \end{array} \\ | \\ \text{CH}_2-\text{O}-\overset{\overset{\text{O}}{\|}}{\text{C}}-\text{R} \end{array}$$

This is actually the result of three consecutive reactions because the fatty acid groups of the triglyceride are rearranged sequentially. Transesterification can be carried out at a relatively low temperature (60 °C) in the presence of an alkaline catalyst (e.g., an alkali-metal alcoxide or hydroxide).

The fat or oil raw material is not a pure triglyceride since significant amounts of other components are also present. Quantitatively, the most significant of these are free (unesterified) fatty acids (1–5 %), which neutralize the transesterification catalyst so that the consumption of catalyst can sometimes be very high. The catalyst dissolves in the joint product glycerol and must be removed as a salt at the end of the process.

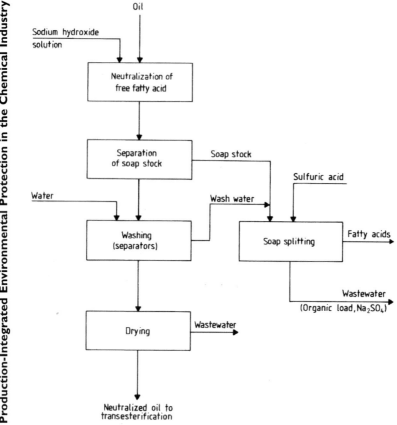

Figure 82. Flow diagram of alkaline refining

Minimization of catalyst consumption is desirable for environmental reasons; therefore, the free fatty acids contained in the oil must be neutralized.

Low-Pressure Transesterification—Old Batch Process. In the old batch transesterification process, neutralized oil is used. The oil is treated with aqueous sodium hydroxide to neutralize the free fatty acids (see Fig. 82). The aqueous phase, which contains a high proportion of the soaps formed (soap stock), is then removed by means of separators, and the remaining soap residues are removed from the oil by washing with water several times (usually twice). These soap residues are then separated [200], [201]; the wash water and the aqueous phase from the first stage, which has a high soap content, are bulked together and treated with sulfuric acid. The soaps are thereby split to form fatty acids and sodium sulfate. The fatty acids can be separated as an organic phase by decantation. The water, which contains sodium sulfate and organic impurities, must be treated in a wastewater treatment plant. The fatty acids produced are a low-value joint product.

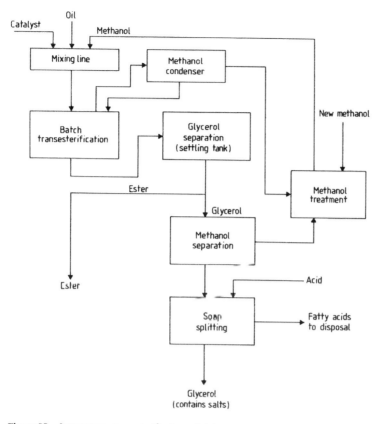

Figure 83. Low-pressure transesterification—batch process

The oil, which is free of fatty acids, is then dried and transesterified (Fig. 83). The batch transesterification process represents the usual state of technology [202]. The process is operated in as simple a way as possible.

Sodium hydroxide dissolved in methanol is used as the catalyst for methanolysis. A high proportion of the sodium hydroxide is converted to sodium methoxide in the solution [203], liberating water:

$$NaOH + CH_3-OH \rightleftharpoons NaOCH_3 + H_2O$$

The dried oil is mixed with methanol and the catalyst solution, and the reaction mixture is pumped into a stirred tank. The reaction is completed there at 80 °C as a batch operation. The methanol that evaporates is completely condensed and returned to the reactor.

At the end of the reaction, excess methanol is partly distilled off. The mixture of fatty acid methyl ester and glycerol is pumped into a settling tank to allow phase separation. After standing for several hours, the lower phase (glycerol phase) is separated. The upper ester phase is removed and temporarily stored before further processing (e.g., hydrogenation to fatty alcohols).

Although excess methanol has been distilled out of the reactor, the glycerol phase still contains methanol, which must be removed before further glycerol processing. Removal of residual methanol is carried out continuously. The glycerol is cooled again and treated with acid in a stirred vessel in a batch process. The sodium soaps formed from the catalyst (see reaction equation), which are almost completely extracted from the mixture by the glycerol,

$$NaOH + CH_3\text{-}O\text{-}\overset{O}{\overset{\|}{C}}\text{-}R \rightarrow NaO\text{-}\overset{O}{\overset{\|}{C}}\text{-}R + CH_3OH$$

are decomposed, forming fatty acid and sodium salt:

$$NaO\text{-}\overset{O}{\overset{\|}{C}}\text{-}R + acid \rightarrow H\text{-}O\text{-}\overset{O}{\overset{\|}{C}}\text{-}R + salt$$

The fatty acids can be separated after a short time as the upper phase.

The old batch transesterification process has the following disadvantages:

1) Transesterification is carried out in only one step, so that a large excess of methanol is necessary to give a high conversion efficiency.
2) The catalyst used is sodium hydroxide dissolved in methanol. Water is formed in the course of alcoxide formation. In the presence of water, the sodium hydroxide present reacts very quickly with fatty acid methyl esters to form sodium soaps. Under these conditions, the soaps formed are not effective catalysts.

For this reason, large quantities of catalyst must be used. This leads to product loss and to large quantities of fatty acids in the crude glycerol after soap splitting. Furthermore, the crude glycerol then has a high salt content. This amount of salt must be removed in subsequent processing and disposed of as waste.

New Method of Low-Pressure Transesterification. The modified process has the following aims:

1) Continuous operation
2) Utilization of the free fatty acids contained in the oil
3) Prevention or minimization of waste products and residues
4) Optimum utilization of the raw materials used, especially reduction in the amount of catalyst and the amount of excess methanol required
5) Heat recovery and optimization of energy consumption

Neutralization of Oils by Pre-Esterification. The free fatty acids present in the raw material are esterified with methanol. The proton-catalyzed esterification can be carried out homogeneously [204]. However, the result is that some of the acidic catalyst is

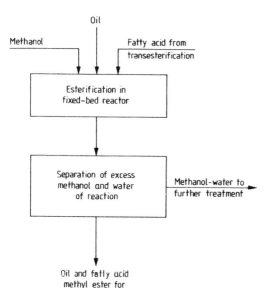

Figure 84. Flow diagram of pre-esterification

also esterified, and the ester and catalyst must then be separated by a water wash. This results in wastewater that sometimes contains toxic substances [205].

To avoid this problem, a heterogeneous catalyst [203], [206]–[208] is used (see Fig. 84) consisting of a strongly acid macroporous ionexchange resin arranged in a fixed-bed reactor. The reaction temperature is between 60 and 80 °C.

In pre-esterification, not only are the fatty acids in the crude oil esterified, but so are the fatty acids produced on splitting the soaps formed in the transesterification process (see Fig. 85). Recycling leads to an increased yield of methyl ester. After pre-esterification, excess methanol is separated and removed together with the reaction water.

The neutralized oil (mixture of triglycerides and methyl esters) is transesterified after addition of methanol and catalyst. Transesterification leads to the formation of glycerol and fatty acid methyl esters. A marked solubility gap exists in the three-component system glycerol–methyl ester–methanol, so that a glycerol phase and an ester phase are formed [208]. This two-phase system gives the possibility of a multistage process. By removing the product glycerol (GLY, see Eq. 1), a high yield of methyl ester (ME) with a reduced consumption of methanol (MeOH) [209] is achieved

$$K = \frac{[\text{ME}]^3 \times [\text{GLY}]}{[\text{MeOH}]^3 \times [\text{TRI}]} \qquad (1)$$

Transesterification is carried out in two stages, the glycerol phase formed being removed after each stage.

The reaction is carried out in tubular reactors with a flow rate that ensures good mixing of the two-phase reaction mixture at the beginning and end of the reaction [210]. Since the amount of back mixing is low in these tubular reactors, short reaction times with low catalyst concentrations can be achieved at ca. 80–90 °C.

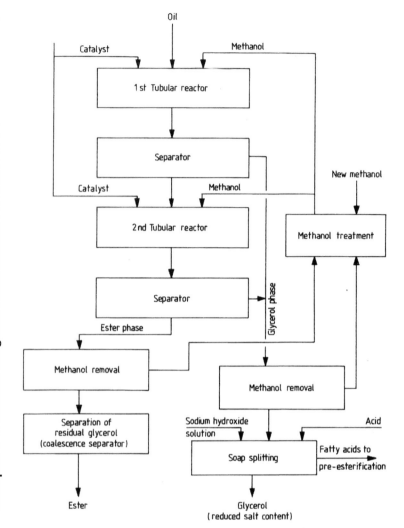

Figure 85. Low-pressure transesterification — continuous process

With a new catalyst (alkali-metal oxide), smaller catalyst concentrations can be used than is the case with sodium hydroxide. With suitable reactor design and choice of catalyst, the salt content of the glycerol produced is decreased by 50%, and smaller quantities of fatty acids are formed in the splitting reaction.

Immediately after transesterification, excess methanol contained in both liquid phases is removed with the aid of a multistage falling-film evaporator. The multistage system enables heat from the hot methyl ester and glycerol to be utilized in the initial evaporation stages.

The recovered methanol at atmospheric pressure is fed to the methanol distillation stage as a vapor.

The soaps in the glycerol from which the methanol has been removed are continuously split with acid. Consumption of acid is minimized by pH control. Concentrated acid is used to minimize the dilution of glycerol with water.

The ester phase, from which the methanol has been removed, still contains undissolved, highly dispersed residual glycerol, which is removed after cooling with the aid of coalescence separators [211]. The use of such equipment avoids the need to wash the esters with water [198]. A highly concentrated crude glycerol is thus obtained, which leads to reduced energy consumption in glycerol processing.

Ecological Aspects of the New Low-Pressure Transesterification Process with Pre-Esterification. In the earlier process, refined oil was used. The waste streams produced in the oil refining process are completely avoided by use of preesterification. Moreover, the waste fatty acid produced in the low-pressure transesterification process can be reutilized.

The reaction is catalyzed by a macroporous cation exchanger. This catalyst has a very long service life and can be disposed of without problems.

By carrying out the transesterification reaction in several stages, the process can be carried out with a lesser excess of methanol. This leads to an energy saving in the methanol recovery and methanol purification processes. The use of a continuous process enables heat to be recovered during the recovery of methanol from the ester and glycerol.

Impure methanol is fed into the methanol purification stage as a vapor, so that the energy consumption in methanol distillation is very low.

The use of an alkali-metal alcoxide in combination with the new reactor design has the advantage of lower catalyst consumption. Since the catalyst is separated along with the by-product glycerol, this means that the amounts of salt produced in a subsequent glycerol processing stage are greatly reduced.

During the process, crude glycerol is diluted as little as possible (use of coalescence separators) to give reduced energy consumption in glycerol processing and to minimize the amount of wastewater produced.

3.2.8. Integrated Environmental Protection in the Production of Vitamins (Example from F. Hoffmann-La Roche)

Introduction. "Etinol" (chemical formula see below) is an important intermediate in the production of vitamins and food dyes, and is produced in a major plant. By using the conventional synthesis route, the raw material consumption of the presented intermediate step is very high, and the by-products formed pollute environmental compartments such as water and air. To produce 1 kg etinol in a multipurpose apparatus, ca. 3 kg raw material is required, so that ca. two-thirds of the raw materials are lost as waste, some in chemically changed form. This situation is unsatisfactory both ecologically and economically.

In principle, two possibilities exist for reducing the amount of waste: to recover the useful raw materials from waste streams and recycle these to the process, or preferably to modify the process so that raw materials are more efficiently utilized.

Original Process for the Production of "Etinol" in a Multipurpose Plant. The reaction consists of the addition of acetylene to a vinyl ketone using lithium as the process auxiliary according to a very simple-looking overall equation:

$$R^1-\overset{O}{\overset{\|}{C}}-CH=CH_2 + HC\equiv CH \xrightarrow[\text{2) }H_2SO_4]{\text{1) Li-NH}_3} R^1-\underset{C\equiv CH}{\overset{OH}{\underset{|}{\overset{|}{C}}}}-CH=CH_2 + Li_2SO_4$$

Vinyl ketone   Acetylene   "Etinol"

However, the reaction must be performed in three complex stages (see below):

Stage I
$$2\,Li + 3\,HC\equiv CH \xrightarrow{NH_3,\ -35°C} 2\,Li-C\equiv CH + H_2C=CH_2$$

Stage II
$$R^1-\overset{O}{\overset{\|}{C}}-CH=CH_2 + Li-C\equiv CH \xrightarrow{NH_3\text{-solvent}} R^1-\underset{C\equiv CH}{\overset{O^-Li^+}{\underset{|}{\overset{|}{C}}}}-CH=CH_2$$

Stage III
$$2\left[R^1-\underset{C\equiv CH}{\overset{O^-Li^+}{\underset{|}{\overset{|}{C}}}}-CH=CH_2\right] + H_2SO_4 \xrightarrow{\text{Solvent}-H_2O} 2\,R^1-\underset{C\equiv CH}{\overset{OH}{\underset{|}{\overset{|}{C}}}}-CH=CH_2 + Li_2SO_4$$

In stage I, metallic lithium is dissolved in liquid ammonia at its boiling point (ca. −33 °C). Acetylene is then fed to this solution, first forming dilithium acetylide ("lithium carbide"). Lithium carbide then reacts further with acetylene to form monolithium acetylide and ethylene. In stage II, the vinyl ketone, diluted with an organic

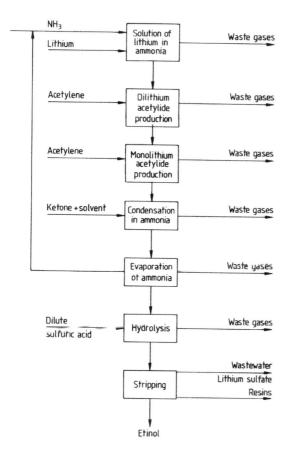

Figure 86. Etinol production—flow diagram of original process

solvent, is added to the lithium acetylide, whereupon "lithium etinolate" is formed. The ammonia that served as a solvent in stages I and II of the reaction is then distilled off from the reaction mixture. In stage III the desired product is obtained from the lithium etinolate by hydrolysis with dilute sulfuric acid.

The former industrial method of carrying out this process is shown schematically in Figure 86. Acetylide formation and reaction of the acetylide with the ketone occurred in that order in the same reaction vessel. Ammonia was then removed in a distillation vessel specially designed for that purpose. The resins produced in stage II of the reaction were removed from the product by a wastewater stripper in the production plant. These resins, along with the wastewater that also contained valuable lithium, were sent to the wastewater treatment plant.

As already mentioned, this process had disadvantages, and great efforts were made to improve it. The production of etinol was described as early as 1948 as the most expensive and difficult part of this vitamin synthesis, and the situation had hardly changed by the early 1980s.

Attempts were made at first to replace the lithium needed in the condensation by cheaper metals such as sodium, potassium, magnesium, or calcium. However, all of these attempts failed because either the yields obtained with the alter-native methods were poor or other disadvantages had to be tolerated, such as the very long times required to dissolve the metals. Attempted improvements to the method, such as recovering the expensive lithium from the wastewater in the form of a salt, were also unsuccessful because of the high resin content of the wastewater.

New Process for the Production of "Etinol" (Fig. 87). *Improvements Developed in the Laboratory.* Laboratory investigations showed that the vinyl ketone used is unstable in liquid ammonia. Therefore, under the conditions of ethynylation in the absence of lithium acetylide or when it was present in only substoichiometric amounts, resins were produced almost exclusively from the vinyl ketone even though the solution was diluted with an organic solvent. Therefore, experiments were made in which the addition was carried out in an organic solvent instead of ammonia. The laboratory results were overwhelming, the yields obtained were 95% instead of 80%, and the lithium excess could be reduced from 20 to 5%. Production of the undesired resins almost ceased [212].

After these laboratory investigations, pilot plant trials were carried out on a modified process. After formation of the monolithium acetylide in stage I, ammonia is distilled off and replaced by solvent I. The distillative removal of ammonia must be carried out at a controlled heating rate so that the monolithium acetylide–ammonia complex, which is not very stable, does not decompose. The distillation is interrupted at exactly 5 °C. In order to facilitate drying of the solvent, a further solvent change is carried out after condensation, since solvent II can easily be dried and is also used in the following step of the process [213].

As a result of the much reduced amounts of resin, a lithium recovery method could be developed. It became apparent that an aqueous solution of lithium hydroxide was especially suitable for this recycling method, provided organic impurities were excluded. The "lithium etinolate" solution is hydrolyzed in stage III with water instead of sulfuric acid. The mixture so produced is in two phases, which are separated in a decantation apparatus. The aqueous solution of lithium hydroxide, which still contains part of the product in solution, is concentrated in a column. The water–product mixture obtained as overhead product is recycled to the hydrolysis stage, and the 10% solution of lithium hydroxide that collects at the bottom of the column is returned to the lithium manufacturer for reprocessing. Only the organic phase is subsequently neutralized with dilute sulfuric acid, and high-boiling components are removed in a stripper column. The consumption of sulfuric acid is considerably lower than in the original process since lithium is obtained not as a salt but as the hydroxide. The amount of sulfuric acid is only that required to neutralize ammonia liberated from the monolithium acetylide–ammonia complex.

Process Changes: Optimization. In the original process, considerable quantities of acetylene and ammonia had to be disposed of as waste products. After the drastic reduction in the amount of resin produced, the consumption of these raw materials could be carefully scrutinized. Improvements can be achieved by:

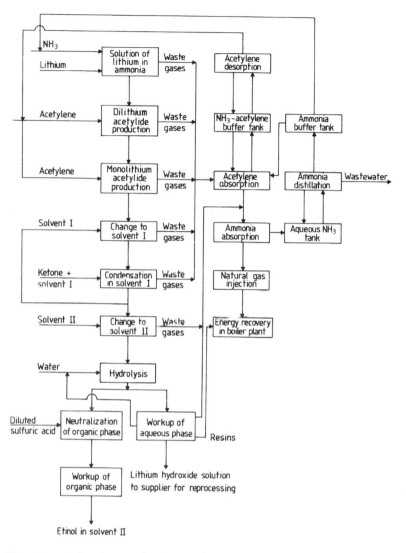

Figure 87. Etinol production—flow diagram of new process

1) Lowering of acetylene consumption by a modified reaction design
2) Recirculation of acetylene and ammonia by means of an absorption–desorption system
3) Utilization of the calorific value of the unused acetylene and of the ethylene produced in the process

The acetylene balance (percentage of total consumption) in the original process is approximately as follows:

Consumption in the reaction:	30 %
Waste gas as ethylene:	20 %
Waste gas as acetylene:	50 %

The important 50 % loss in the form of acetylene can be recovered and returned to the reaction cycle. All of the waste gases from the steps up to the condensation stage are fed to the central waste-gas recovery system. The gases are first passed through the acetylene absorption plant where most of the acetylene is removed from the waste-gas stream by supercooled ammonia with a low acetylene content. Because the other components of the waste gas are less soluble than acetylene in liquid ammonia they pass through the absorption column. The acetylene-rich ammonia from the bottom of the column goes to a central ammonia–acetylene buffer tank. The gases leaving the acetylene absorption column, together with waste gases from the neutralization stage and from all processes of stage III, are passed through the ammonia absorption column, where the waste-gas stream saturated with ammonia is scrubbed with water. The aqueous ammonia solution in the lower part of the column is retained in the absorption recirculation system long enough for the concentration of ammonia to reach ca. 15 %. The ammonia–water mixture is separated by a pressurized rectification column [214]. The remaining waste gases after ammonia absorption—mainly ethylene and nitrogen with small amounts of acetylene—are fed to the boiler plant for burning as fuel.

Acetylene recovered by the absorption process is reused for lithium acetylide production in stage I of the process. For this, the acetylene-containing ammonia from the central ammonia–acetylene buffer tank is fed into a desorption column where the acetylene is driven off from the ammonia by heating and is obtained as overhead product. Ammonia from the bottom of the column is reused for dissolving metallic lithium or absorbing acetylene.

The idea of burning the high-energy waste gases acetylene and ethylene in an environmentally friendly process to produce steam is certainly attractive. However, serious safety problems must be solved before such a process can be considered. Particular attention must be paid to the safety aspects of acetylene, which is explosive over a wide range of concentrations in air (lower flammability limit 2.3 vol%, upper flammability limit 82 vol%) and which can decompose at moderately elevated pressure even in the absence of oxygen, generating large amounts of heat and high pressure. Investigations have shown that self-decomposition can be suppressed by diluting the acetylene with natural gas, even when the gas mixture is ignited. Since the boiler plant runs on natural gas, this principle has been used in combustion of the waste gas. Waste gases from the production plant are diluted to 40 % by injection of a stream of natural gas after they have passed through the absorber; they are then compressed to the pressure required by the boiler plant. Moreover, the production plant itself is separated from the combustion plant by a system of flame arrestors [215], [216].

Energy Considerations. Both the process of dissolving metallic lithium and the condensation of lithium acetylide with ketone are exothermic. Since these reactions are carried out at the boiling point of ammonia (–33 °C), or in any case below 0 °C, a cooling system for very low temperatures is inevitable. The cooling medium used is identical to the reaction solvent (in this case, ammonia). Thus, not only can cooling be carried out without using chlorofluorohydrocarbons, but also any accidental mixing of the cooling ammonia with the ammonia of the process will not lead to problems, since it is identical with the solvent used. The pressure difference guarantees that in case of a leak the only possible flow will be of coolant into the reaction solution. Also, the strong smell of ammonia facilitates leak detection, so that amounts of coolant lost by leakage are very small.

The refrigeration process uses compressors and an expander system. The proximity of the refrigeration machinery enables the compressed hot ammonia to be used as a source of heat in the etinol plant. Thus, the acetylene desorption column used to drive off dissolved acetylene is heated by waste heat from the refrigeration machines. Only the excess heat is removed by a cooling tower, and this requires a minimum amount of cooling water.

Ethylene and acetylene are diluted with natural gas in the gas injection equipment and fed to the boiler plant to utilize the energy they contain. The resins can also be used for energy production with the aid of suitable incineration plants. Thus, the high calorific values of these by-products and wastes can be used to produce energy in the form of steam and electricity, thereby avoiding the environmental pollution that would result from the burning of fossil fuels.

Environmental Protection Balance. The material flow diagrams for the old process and for the new improved process are shown in Figure 88.

The presence of lithium salt in the wastewater is avoided. Also, the waste-air streams conform with legal requirements and are all used either in the process or for energy production. Increased recycling of ammonia and acetylene leads to reduction in the consumption of ammonia to 25 % of its former value; acetylene consumption is reduced by 50 %. Thus, the stoichiometric excess of acetylene used in stage I of the reaction is reduced to ca. 12 %.

The increased extent of closed material cycles is remarkable. Not only ammonia and acetylene, but also lithium in the form of lithium hydroxide solution, are recycled. The efforts toward a better environmental compatibility have resulted in a new process that is more complex than the original "straight-through" process. The effects are beneficial for both the ecology and the economy. The performance of new installations is optimized by ensuring that individual process steps harmonize with each other. This applies not only within the identical batch plants used for the different stages, but also to energy utilization in the boiler plant. Operators of these plants are required to have an increased level of understanding. Improved training and education can ensure that consumption of materials is minimized and utilization of equipment and raw materials is maximized.

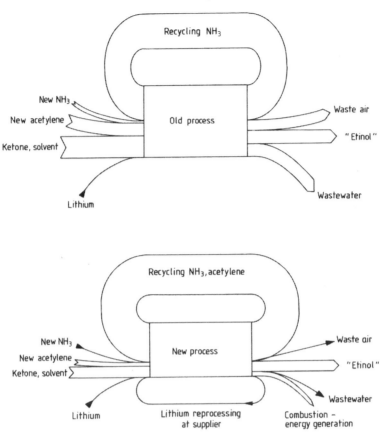

Figure 88. Comparison of mass flow of old and new processes

3.2.9. Production of Pure Naphthalene without Residues—Replacement of Chemical Purification by Optimized Multiple Crystallization (Example from VFT)

Formation and Uses of Naphthalene. In the production of coke for use in blast furnaces, coal tar is produced as a liquid by-product. Raw materials for the chemical industry, metallurgy, and energy production are obtained from this complex mixture of chemical compounds with aromatic character [217].

The most important constituent of coal tar with respect to quantity is naphthalene, which is present at a concentration of 10 %. Pure naphthalene (up to 99.9 %), obtained by a refining process, is of great economic importance as a basic chemical for the

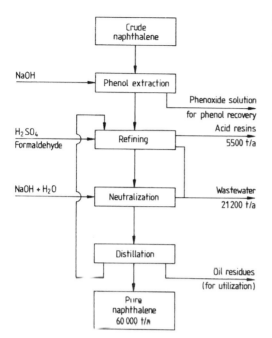

Figure 89. Flow diagram of old process of chemical naphthalene refining by treatment with sulfuric acid and sodium hydroxide solution

production of, e.g., phthalic anhydride-based plasticizers, dyes, plant-protection agents, and special solvents [218].

Chemical Refining. The conventional method of purifying naphthalene uses several stages of physical and chemical treatment:

1) Coal tar distillation
2) Crystallization of the naphthalene fraction
3) Treatment of the crystallized product with concentrated sulfuric acid and formaldehyde, followed by neutralization with sodium hydroxide solution
4) Redistillation to remove dark-colored secondary constituents (Fig. 89)

This refining process also produces highviscosity naphthalene-containing acidic resins amounting to ca. 8% of crude naphthalene. These must be disposed of by dumping.

Also, based on the original weight of coal tar, naphthalene-containing waste sulfuric acid (7%) and naphthalene-containing dilute sodium hydroxide solution (ca. 18%) are produced. These waste streams can be neutralized by mixing them together and are then sent to a sewage plant. However, because the material has large amounts of naphthalene-containing oils, sludge deposits are formed that require very complex treatment.

Physical Refining Process (Solution of the Environmental Problem). VFT realized the advantage of preventing the production of waste materials by removing the impurities in naphthalene simply by repeated application of a physical separation

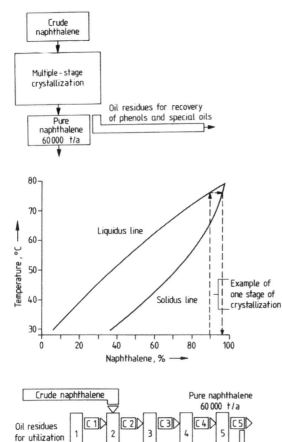

Figure 90. Flow diagram of new process of physical refining by countercurrent crystallization

Figure 91. Phase diagram of benzo[*b*]thiophene – naphthalene

Figure 92. Schematic of five-stage crystallization

process. High selectivity is required to enable the residual oil to be used as a raw material in later process stages without further treatment (Fig. 90).

To achieve this desirable situation, a new process had to be developed. The phase diagram of naphthalene and the main impurity benzo[*b*]thiophene is given in Figure 91. It shows that with a starting mixture of 90 % naphthalene, a single-stage crystallization gives only a 6 % increase in concentration with a very low yield. On the other hand, a multistage crystallization process can give naphthalene in good yields and high purities that meet market requirements (Fig. 92).

Process Details and Plant Technology. As an illustrative example, a detailed description of stage 2 of the process is given. The crude naphthalene, together with the residual oil from stage 3 and the crystalline product (C 1) from stage 1, is cooled in the crystallizer at controlled temperature. This causes a layer of crystals with an increased naphthalene concentration (in accordance with the phase equilibrium) to form on the cooled surface of the inner wall of the crystallizer.

The depleted residual melt is recycled to stage 1, and is there separated into a residual oil and the crystallized product (C 1). Benzo[*b*]thiophene and other impurities in the crude naphthalene remain in the residual oil, which can be processed in adjoining production plants.

Naphthalene shows a marked tendency to undergo heterogeneous seed formation on cooling of the melt [219]. A crystallizer was therefore developed in which the molten naphthalene fraction runs down the interior of vertical tubes as a falling film [220]. During the crystallization phase, the depleted naphthalene oil that runs off the inside of the tube is collected in a vessel and recycled by a pump to the falling-film distributor at the top of the crystallizer. The outside surfaces of the crystallization tubes are covered with a falling film of water, which acts as a cooling or heating medium. The temperature of the water is lowered (crystallization phase) or raised (melting phase) in accordance with a predetermined function.

At the end of the crystallization phase, the liquid residual oil is pumped off. In the next stage of the process, the falling film of water is heated under temperature-controlled conditions so as to give partial melting of the crystal layer on the inside of the tube. This causes melting of that fraction of material that contains a higher proportion of impurities (in accordance with the phase diagram). These fractions are collected separately in the vessel under the crystallization tubes and are distributed to various crystallization stages according to their purity, as shown in Figure 92.

A crystallizer containing >1000 tubes, all 12 m in length, was designed and built to give an economically viable production capacity. Careful attention was given to the design of the crystallizer, the equipment for providing heat energy at the correct rate, and precise process control.

This multistage crystallization process is controlled by a central control system. In this way, the energy and mass flow rates can be continuously balanced in the various crystallization stages. Determination of the process parameters and control of the production program are based on the analytically determined starting concentrations at each stage and the desired separation factors.

Since crystallization is a purely physical separation process, neither solid waste nor wastewater is produced. The comparatively small flows of waste air due only to changes in the contents of the plant are taken to the central waste-gas combustion plant. All impurities in the crude naphthalene are concentrated in the residual oil from stage 1 and can be economically utilized in further processing stages. This naphthalene purification process was installed in the Castrop-Rauxel works of VFT, and has produced > 60 000 t/a of pure naphthalene consistently and without problems for more than 10 years.

Benefits of the Plant. The shortage of dumping space has led to a drastic increase in dumping costs, so that production of pure naphthalene by chemical refining is no longer economically possible. Development of this residue-free environmentally friendly process has made a decisive contribution to the stabilization of coal tar utilization in Germany.

3.2.10. Improvements in the Polypropylene Production Process (Example from Shell)

Introduction. The discovery by ZIEGLER and NATTA of the catalyst system named after them in the 1950s opened the way to the large scale production of polypropylene. Shell companies were among the first to realize the potential of this process and started a development program that culminated in the start-up of Shell's first two polypropylene units in Carrington, United Kingdom (1962) and Pernis, The Netherlands (1963). These plants were very small (capacity 5×10^3 t/a) compared to present-day standards and used first-generation Ziegler–Natta catalysts. These catalysts had a low activity and a low selectivity, making it necessary to remove the catalyst residues as well as the atactic polymer by de-ashing. This process made use of a hydrocarbon diluent, which resulted in relatively low system pressures.

In 1964, Shell Research discovered a catalyst with improved selectivity, which led to a significant rationalization of the production process [221]. The removal of atactic polymer could be omitted because polymer of the desired isotacticity could be made directly in the reactor. The above developments formed the basis for a continuous improvement in plant design, with production capacities steadily rising to 100×10^3 t/a in the late 1970s.

Major Developments since 1980. During the 1980s, two fundamental changes in the polypropylene process led to a breakthrough in terms of improved process economics, energy efficiency, and reduced environmental impact. The first of these changes was the conversion from a diluent process to a process in which liquid propene acts as both reactant and "diluent" for the reactor slurry. This mode of operation, which is called LIPP (Liquid Propylene Process) leads to a considerable increase in the pressure at which such a process is operated.

The second breakthrough was the use of a high-activity catalyst. This proprietary catalyst called "Shell High Activity Catalyst" (SHAC) was developed by Shell Research and achieved a hundredfold improvement in product yield with respect to the quantity of titanium catalyst employed, leading to a titanium residue in the polymer on the order of 1 to 2 µg/g [222]. As a result, removal of the catalyst residues could be omitted, leading to further process simplification. An impression of the simplification in terms of unit operations between the "diluent" process and new "LIPP" process can be gained from Figures 93 and 94.

Capital Cost. Clearly, from the above, the investment cost per tonne of capacity for the LIPP–SHAC process will be appreciably lower because fewer equipment items are required in the streamlined process.

Energy Consumption. Without the need for de-ashing and diluent distillation, and with a more efficient operation due to the higher pressure level in the reactor, major reductions (ca. 50% expressed in terms of fuel equivalent per tonne of polymer) in the energy consumption of the modern polypropylene plant have been achieved. Figure 95

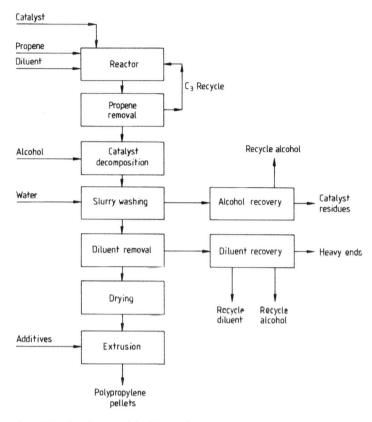

Figure 93. Flow diagram of the diluent–slurry process

shows the energy consumption per tonne of polypropylene on changing from a diluent to a LIPP process.

Environmental Impact. With regard to environmental impact, four areas can be identified in which clear progress is evident:

1) *Metal salts* The metal salts that originated from the de-ashing step required disposal. Since de-ashing is no longer needed, their quantity has been diminished by a factor of >100. The only source of metal salts remaining is equipment flushing where small amounts are found.
2) *Diluent* Although extensive precautions are taken to avoid loss of diluent from the plant, this can never be totally prevented. Apart from the fugitive losses of diluent, a loss occurs in the heavy ends fraction of the diluent recovery operation. Since the LIPP process does not use a diluent except for some flushing, the amount of diluent lost is drastically reduced.
3) *Alcohol* In the diluent process, an alcohol is used to deactivate the catalysts. Although the alcohol is recovered from the aqueous stream, a small residual

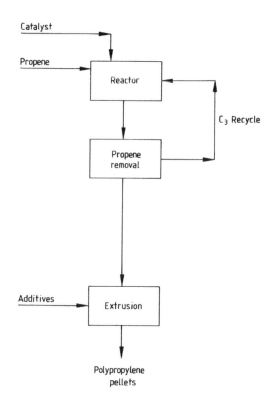

Figure 94. Flow diagram of the bulk polypropylene process (LIPP)

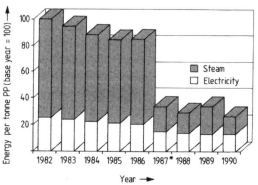

Figure 95. Energy consumption of polypropylene (PP) plants
* Conversion to LIPP process, 1987.

amount will be contained in the wastewater. Since no alcohol is required in the new process, this discharge no longer exists.

4) *Propene* The overall streamlining of the process has resulted in a considerable reduction in fugitive emission of propene from the plant.

The relative improvements in the above-mentioned fields are summarized in Figure 96.

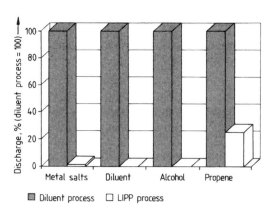

Figure 96. Environmental discharges from polypropylene production processes

Future Developments. Current developments are mainly in the field of catalyst improvement, both of conventional Ziegler–Natta catalysts and of single-site catalyst. These developments will mainly influence the properties of the polymer product rather than the manufacturing process.

3.2.11. The Zero-Residue Refinery Using the Shell Gasification Process (Shell–Lurgi Example)

Introduction. Refineries worldwide are being subjected to increasing pressures, both legislatively and economically. The effects of the Clean Air Act (CAA) in the United States of America are well documented [223]. In Europe, increasing environmental legislation is making itself felt both in refinery operation and in terms of product quality. The former is governed by the requirements of the EC Large Combustion Plant Directive or more stringent national regulations. The latter is dictated by EC Directive EN 228 requiring a reduction of sulfur in diesel motor fuels to 0.05 % S by 1996 and in heating oils to 0.1 % S by 1999.

Simultaneously, a medium-term pressure is occurring to reduce the low-sulfur crude intake of refineries because of declining availability and an increasing cost differential compared with high-sulfur crudes.

Various schemes including either coking or residue gasification have been proposed to cope with these pressures. Realization that coke disposal is becoming increasingly difficult has led to wider application of residue gasification. Four existing refineries in Germany already employ the Shell Gasification Process (SGP) for residue utilization. Worldwide, five plants are currently under construction, including a major revamp at Shell's own Pernis refinery near Rotterdam.

SGP in the Refinery Flow Sheet. Before describing the gasification and gas treating technology, the way they fit into the overall picture of the refinery flow sheet should be

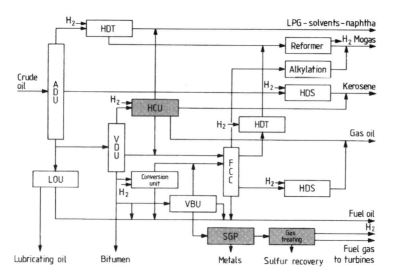

Figure 97. Shell gasification process in the refinery environment
ADU = Atmospheric distillation unit; LOU = Lube oil unit; HDT = Hydrotreating; VDU = Vacuum distillation unit; HCU = Hydrocracker unit; VBU = Visbreaker unit; FCC = Fluid catalytic cracking; HDS = Hydrodesulfurization; SGP = Shell gasification process; Mogas = motor gasoline; LPG = Liquefied petroleum gas

reviewed. Figure 97 provides a simplified refinery flow sheet after implementation of a combined hydrocracker (HCU) and SGP project. These two units, together with the gas-treating section downstream of the SGP, are shaded in the diagram.

The main environmental contaminants, sulfur and heavy metals, tend to remain in the bottom product of distillation processes. Their path through the refinery is from the crude oil intake via the atmospheric distillation unit (ADU), through the vacuum distillation (VDU) to the visbreaker unit (VBU), a mild thermal cracking process with subsequent distillation facility. In the original flow scheme, prior to implementing the new units all the residue from the visbreaker together with its load of contaminants, typically 2–4% sulfur and up to 400 µg/g vanadium, was sold into the heavy fuel oil pool. This material is largely used in oil-fired power stations, where the restrictions of the Large Combustion Plant Directive are now placing restraints on its use, particularly to reduce atmospheric sulfur emissions.

Reductions in the allowable sulfur content of diesel and heating oil both require the use of more hydrogen in the hydrodesulfurization (HDS) and hydrotreating (HDT) units in the gas oil fractions. The new hydrocracker, whose main purpose is to convert middle distillates into lighter fractions, is also a large consumer of hydrogen.

Reduction in allowable benzene levels in motor gasoline (Mogas), which the CAA requires and which may yet become a requirement in Western Europe, requires a reduction of operating severity on the reformer, which reduces the amount of by-product hydrogen available from this unit. Environmental pressures as expressed in legislation to reduce sulfur emissions to the environment from refinery products, especially when combined with a reduction of potentially carcinogenic material from gasoline, have the following results:

1) A surplus of high-sulfur fuel oil
2) A large increase in the internal hydrogen demand of the refinery
3) A substantial decrease in the amount of internally generated hydrogen from the reformer

The overall technological concept provides for utilizing the surplus high-sulfur fuel oil to manufacture the hydrogen required to cover the ever-increasing internal hydrogen deficit. It in cludes the construction of a new hydrocracker and the residue gasification unit for the production of hydrogen and sulfur-free fuel gas, which will be fed to a gas turbine that forms the heart of a combined-cycle cogeneration plant. Additionally, it includes new amine treating and Claus units.

The feedstock for the SGP unit is a heavy, vacuum-flashed cracked residue from an existing visbreaker unit. This material is typically characterized by high viscosity and high metal content.

The product hydrogen would be aimed primarily at supplying the hydrocracker. Sufficient flexibility can, however, be built into the gas treatment facilities to accommodate alternative scenarios in the overall hydrogen refinery balance and to produce methanol for methyl-*tert*-butyl ether (MTBE) production or other uses.

Process Description. *General.* When looking at the market possibilities, the sales volume of the product should be matched with the total volume of residue available. Of those products of immediate use in the refinery that can be produced from residues, hydrogen is an obvious choice. Advanced processing schemes consume large quantities of hydrogen. Whether the hydrogen demand will match the quantities of residue available will of course depend on overall refinery configuration. Methanol, either as a direct fuel additive or as a feedstock for MTBE production, can account for utilization of some additional residues. Ammonia prices are too low to make it attractive, and there is no marketing synergy as might be the case with methanol. Speciality chemicals do not generate sufficient volumes to make any impact on the residue problem, although a sidestream unit on a base plant might offer possibilities for a high margin investment. The market for electrical power is acquiring increasing importance as another possibility for absorbing the quantities of residue available [224].

The block flow diagram (Fig. 98) illustrates the plant arrangement for the simultaneous production of all three main products indicated above — power, hydrogen, and methanol — together with the possibility of manufacturing pure carbon monoxide as well. Although, in practice, a plant would rarely be required for all four products, two product schemes are by no means uncommon.

Gas Production. The feedstock is gasified in the SGP reactor (a) with oxygen to produce raw synthesis gas at about 1300 °C and 60 bar (Fig. 99). This gas is a mixture of hydrogen and carbon monoxide, which also contains CO_2, H_2S, and COS together with some free carbon and ash from the feedstock.

Recovery of the sensible heat in this gas is an integral feature of the SGP process. Primary heat recovery takes place in a waste-heat exchanger (b) generating high-

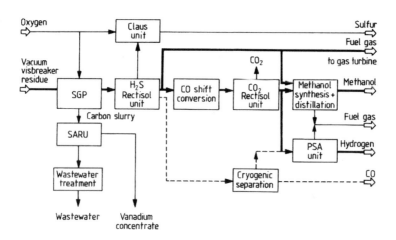

Figure 98. Block flow diagram of an SGP-based methanol and hydrogen plant
SGP = Shell gasification process; SARU = Soot ash removal unit; PSA = pressure swing adsorption

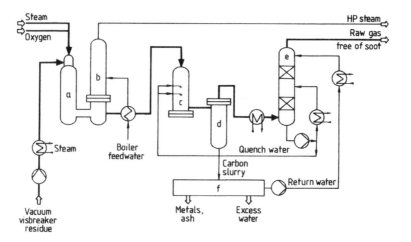

Figure 99. Flow diagram of an SGP unit based on residual oil
a) Reactor; b) Waste-heat exchanger; c) Quench pipe; d) Carbon separator; e) Scrubber; f) Soot ash removal

pressure saturated steam in which the reactor effluent is cooled to about 340 °C. The waste-heat exchanger is of a special design developed specifically for these operating conditions. Part of the steam thus generated is used for feedstock and oxidant preheating; the remainder is superheated for use in the CO shift conversion and in steam turbine drives. Steam conditions are chosen to be optimized with the steam turbine part of the combined cycle.

Particulates are removed from the gas by means of a two-stage water wash. Carbon formed in the partial oxidation reactor is removed from the system as a carbon slurry together with the ash and the process condensate. This slurry is subsequently processed in the ash removal unit (f) described below. The product synthesis gas leaves the scrubber at a temperature of about 40 °C and is essentially free of carbon. It is then suitable for treatment with any commercial desulfurization solvent.

Gas Treatment. Selection of the actual process to be used for desulfurization depends on the application and on site-specific conditions. The very high purity required for methanol synthesis gas (S < 0.1 ppm) practically mandates the Rectisol process, which uses cold methanol as a solvent. For power applications, the high H_2S–CO_2 selectivity of the Purisol solvent *N*-methyl-2-pyrrolidone (NMP) makes it a good choice. Where low investment costs have priority over reduced operating costs methyldiethanolamine (MDEA) may be used. In all cases, the desulfurized gas is suitable for firing in a gas turbine as part of an integrated gasification combined cycle (IGCC) power plant.

The remaining gas is then subjected to CO shift conversion using steam to convert the CO to CO_2 and hydrogen. Hydrogen purification to > 99 % can be effected by pressure swing adsorption (PSA) as shown in Figure 98, either with or without prior CO_2 removal. An alternative scheme of CO_2 removal followed by methanation produces a 98 % pure H_2 product with < 10 ppm carbon oxides.

A side stream of gas is bypassed around the shift unit to achieve the correct $H_2 \cdot CO$ ratio for the methanol synthesis process. Methanol synthesis itself operates at about 80 bar using the Lurgi Low Pressure Methanol Process followed by a distillation step to achieve the required product quality specification.

The provision for carbon monoxide production is shown by the dotted line in Figure 98. CO_2 is removed to a greater extent than for hydrogen or methanol production before the gas enters the cold box of the cryogenic unit. This produces a CO product that meets the specification for acetic acid production and a > 97 % hydrogen stream at about 22 bar. Depending on the overall hydrogen scheme, this may be recompressed and fed through the PSA for final cleanup or used raw in the refinery.

Side-Stream Units. Wash water from particulate removal is treated in the soot ash removal unit described in more detail below. Carbon and ash are separated from the water by filtration. The filter cake is worked up by controlled oxidation to a salable vanadium concentrate. The bulk of the water is recycled to the water wash.

When using a physical wash such as Rectisol or Purisol, the regeneration system of the desulfurization solvent can be designed to operate selectively so that it produces a gas of sufficiently high H_2S content to allow it to be processed to elemental sulfur in the sulfur recovery unit. In many instances, the Lurgi Oxygen Claus technology can be used at this location, either to de-bottleneck existing units or for new plants.

Process Refinements. Handling the carbon slurry is an important aspect of partial oxidation processes. The traditional approach to handling the carbon slurry was to contact it with a hydrocarbon to retain the carbon, which could thus be recycled to the reactor. The early Shell units used fuel oil as the extraction medium. This allowed atmospheric operation, providing a plant with low investment costs and, at least initially, low operating costs as well. The economics of this approach deteriorated with heavier feedstocks. The pelletizing oil had to be purchased at a premium to the main

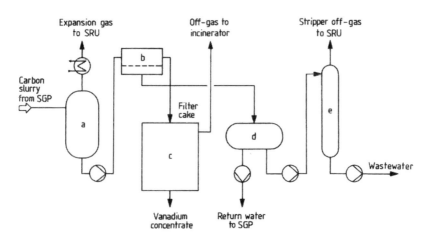

Figure 100. Flow diagram of a soot ash removal unit
a) Slurry tank; b) Filter; c) Multiple-hearth furnace; d) Return water vessel; e) Wastewater stripper
SGP = Shell gasification process; SRU = Soot removal unit

feedstock, and with higher ash contents, 100% carbon recycle became nearly impossible since much of the ash was recycled as well.

The logical development of the process was to use naphtha as extraction medium. Soot is extracted from water with naphtha in the form of pellets. These naphtha–soot pellets are sieved off from the water and then mixed in with the main feedstock at whatever temperature is required to achieve the desired viscosity. The naphtha is then distilled off from the feed and recycled to the extraction stage leaving the soot behind in the feed. The use of naphtha as an intermediate allowed the use of heavier, more viscous feedstocks than in the case of pelletizing with fuel oil. Also, improvement in the separation of carbon and ash allowed 100% recycle. Nonetheless, an ash buildup factor of about 3:1 can be observed under 100% carbon recycle conditions. The plant is costlier than fuel oil pelletizing in both investment and operation.

These facts have led to the review and finally the development of an alternative approach and in particular the possibilities of recovering vanadium for metallurgical use. Figure 100 illustrates the principles of the soot ash removal unit. The carbon slurry from the SGP unit is flashed to atmospheric pressure in the slurry tank (a). The slurry is then filtered on an automatic filter (b) to recover a filter cake with about 80% residual moisture and a clear water filtrate. The filter cake is subjected to a controlled oxidation process in a multiple-hearth furnace (c). This type of furnace, which is well established in many industries and specifically in the vanadium industry, allows combustion of the carbon to occur under conditions where the vanadium oxides neither melt nor corrode. This is not an easy task if one considers the problems of burning a high-vanadium fuel oil in a conventional boiler. The product is a vanadium concentrate, which contains about 75% V_2O_5. Compared to the old naphtha extraction-based recycle system, the new once-through process consists of only two proces-

Table 5. Refinery mass balance, 10^6 t/a

Mass balance	Before 1993	After 1997
Feedstock		
High-sulfur crude	11.3	15.0
Low-sulfur crude	4.5	2.8
Other feedstocks	2.1	0.1
Total intake	17.9	17.9
Production		
White product make	12.8	14.0
Fuel oil make	3.5	2.1
Sulfur	0.11	0.30
Total output	16.4	16.4

Table 6. Refinery and global emissions, 10^3 t/a

Emissions	Before 1993	After 1997
Refinery emissions		
SO_2	35	24
NO_x	12	7
Particulates	4.0	1.9
CO_2	5 000	6 100
Global emissions (refinery + products)		
SO_2	348	206
CO_2	56 600	56 800

sing steps, which are not integrated with the gasification section of the unit. It is cheaper both in investment and in operating costs.

An additional benefit of this development is increased feedstock flexibility, both for the SGP itself and for the refinery as a whole. The carbon recycle currently practised also recycles some of the ash, so that the charge pump, burner, and reactor system are exposed to a higher content of ash than that present in the fresh feed. Elimination of the recycle thus makes it possible to use feeds with considerably higher ash contents than previously. Current experience of >1000 µg/g vanadium at the reactor inlet would be directly applicable to residues of this quality. This is certainly an important feature when reviewing the possibility of introducing heavier (and cheaper) crudes into the refinery.

Environmental Aspects. This article has concentrated on the SGP as part of an overall refinery modernization project. Evaluation of the environmental impact can, however, be performed only in the context of such a plant. The data in Tables 5 and 6 are taken from [225] which describes the project under construction in the Shell refinery in The Netherlands.

The refinery mass balance of Table 5 shows a decisive shift to high-sulfur crude in the future increasing the total intake of high-sulfur crudes from 63 to 84 % of total feed. Nonetheless, the refinery SO_2 emissions and the sulfur content of products are both reduced by one-third, as can be seen in Table 6. Similar reductions can be seen for NO_x

and particulate emissons. CO_2 emissions from the refinery increase, but because of the lighter product slate, global CO_2 emissions are hardly affected. All this is achieved with an increase in the high-value white product from 78 to 85 % of total output and a reduction of 40 % in fuel oil output.

The technology described has been applied in a number of refineries as a means of manufacturing chemicals from low-value feedstocks. With increasing environmental awareness and continued refinement of the technology, it is now finding a strategic place in the planning of low- or even zero-residue refineries [219]–[227a].

3.2.12. Neutral Salt Splitting with the Use of Hydrogen Depolarized Anodes (Hydrina-Technology, Example from De Nora Permelec)

General. Worldwide, more than 3×10^6 t of sodium sulfate is obtained per year as a by-product of many industrial processes: only a portion of this production can be commercialized after crystallization and calcination, e.g., for the detergent and glass industries, which, however, have been progressively reducing their consumption in the last ten years. The remaining portion of sodium sulfate is usually diluted and led into the bulk of effluents or disposed of.

However, disposal will become increasingly difficult, if not impossible, in the near future due to tighter environmental regulations. In any case, it will certainly become increasingly expensive and for this reason, at least, the recovery by electrolysis of sodium sulfate—and in general of other neutral salts, such as sodium nitrate and sodium chloride, which are also generated in some chemical processes—is expected to have a bright future.

The overall economics of the recovery of by-product salts by electrolysis benefits from the saving of the disposal cost as well as from the production of caustic and acid, which may be recycled to upstream plants. The success of this operation depends on a number of conditions including

1) Substantial savings of money, which otherwise would have been required for disposal of by-product salts. In the specific case of sodium sulfate, electrolysis becomes economically interesting when the disposal cost of the salt is in the order of $ 100 per tonne, with a caustic soda price in the range of $ 200 per tonne.
2) Electric energy consumption per tonne of produced caustic soda, which is close to that of membrane chlor–alkali electrolysis (i.e., 2200–2500 kW · h/t).
3) Simple and reliable integration with the plants where the by-product salts are generated. This integration may require auxiliary sections designed to eliminate impurities, to concentrate by-product salt solutions when these are excessively dilute, (e.g., <10 wt %), and, in some cases, to concentrate the acid produced.
4) Negligible requirements of manpower for both operation and maintenance.

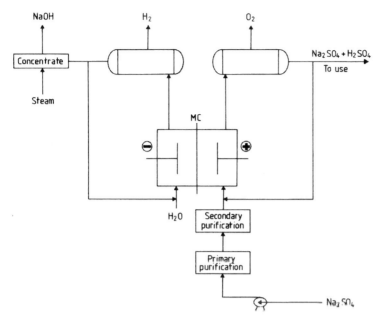

Figure 101. Flow diagram of sodium sulfate electrolysis in two-compartment electrolyzers (production of acid sodium sulfate, oxygen-evolving anodes)
MC = Cation-exchange membrane

Structure of the Electrolyzers. The following discussion is based on sodium sulfate, which is the by-product salt of greatest concern [228], [229]. However, many of the conclusions also apply to the electrolysis of other salts such as potassium sulfate and sodium and potassium nitrate.

Industrial electrolyzers are made of rows of electrically connected elementary cells, preferably of the bipolar type. Many elementary cell structures have been described in the literature. The most important from the industrial point of view are:

1) Elementary cells comprising two compartments separated by a cation-exchange membrane (MC), as illustrated in Figure 101, to produce caustic soda and a mixture of sulfuric acid and unconverted sodium sulfate, hydrogen, and oxygen.
2) Elementary cells comprising three compartments separated by two membranes, namely a cation-exchange (MC) and an anionexchange (MA) membrane (see Fig. 102). In this case the electrolysis products are caustic soda, sulfuric acid, hydrogen, and oxygen.
3) Elementary cells comprising three compartments separated by two cation-exchange membranes as illustrated in Figure 103. In this case, the products are again caustic soda, a mixture of sulfuric acid and unconverted sodium sulfate, hydrogen, and oxygen, as in the case of the two-compartment process. The additional anodic compartment acts as a buffer section so that the solution to be electrolyzed is separated from the anode. This scheme is particularly useful to prevent:

143

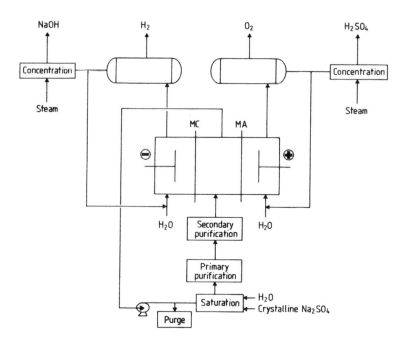

Figure 102. Flow diagram of sodium sulfate electrolysis in three-compartment electrolyzers (production of pure sulfuric acid, oxygen-evolving anodes)
MC = Cation-exchange membrane; MA = Anion-exchange membrane

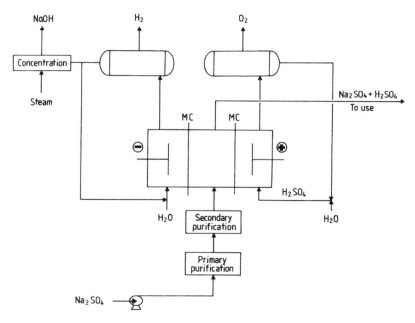

Figure 103. Flow diagram of sodium sulfate electrolysis in three-compartment electrolyzers (buffer compartment)
MC = Cation-exchange membrane

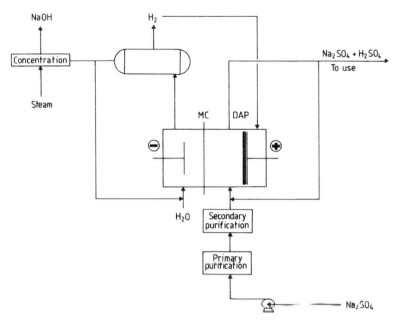

Figure 104. Production of acid sodium sulfate in the Hydrina membrane electrolyzer
MC = Cation-exchange membrane; DAP = Depolarized anode package

- Anodic side reactions, as in the presence of certain oxidizable anions, such as chlorate [230]
- Corrosion of the anodes due to the high concentration of sulfuric acid and unconverted sodium sulfate at elevated temperature
- Decay of the membranes caused by large contents of impurities in the feed solutions [231]

A problem affecting all three schemes is the high cell voltage, which involves high energy consumption. The high cell voltage results from the energy required to perform the following reaction at an industrially useful rate:

$Na_2SO_4 + 3 H_2O \rightarrow 2 NaOH + H_2SO_4 + H_2 + 0.5 O_2$

Energy consumption may be substantially reduced by replacing the oxygen-evolving anode with a hydrogen depolarized anode [230], [232]. The modified schemes are shown in Figures 104 and 105.

The energy consumption is decreased remarkably according to the following new overall reaction:

$Na_2SO_4 + 2 H_2O \rightarrow 2 NaOH + H_2SO_4$

Figures 104 and 105 demonstrate that the hydrogen consumed at the depolarized anode is the same hydrogen evolved at the corresponding cathode. This hydrogen need only be washed in a suitable scrubber to eliminate entrained catholyte mists. The

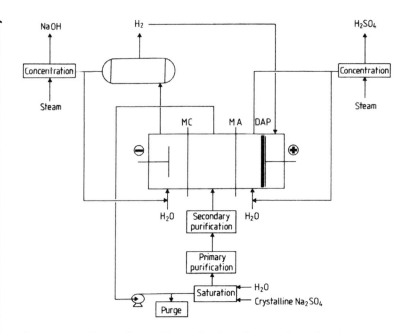

Figure 105. Production of pure sulfuric acid in the Hydrina membrane electrolyzer
MC = Cation-exchange membrane; MA = Anion-exchange membrane; DAP = Depolarized anode package

hydrogen thus obtained has a high purity, which allows for long-term stability of the catalyst of the depolarized anode.

A conventional structure used for the hydrogen depolarized anode consists of an electrically conductive sheet with controlled porosity and hydrophobicity, coated with a catalytic layer. Under optimum operating conditions the electrolyte partially fills the pores, with the gas–liquid meniscus in the area of the sheet containing the catalyst.

Operating conditions causing percolation of the electrolyte through the pores or loss of gas in the electrolyte should be avoided. For this reason, anodes of this type allow rather small pressure differentials between the electrolytic compartment and the hydrogen chamber. The height of these anodes is thus limited. Thus, cell designs with segmented compartments have been proposed [233]. Besides this limitation, conventional electrodes are affected by problems caused by the presence of electrolyte in the pores in direct contact with catalyst. Therefore, both crystallization phenomena are experienced, which result in destruction of the porous sheet structure and poisoning of the catalyst when the electrolyte contains heavy metals, even in trace amounts.

These problems are overcome by using a depolarized anode made of a porous sheet containing a catalyst, bonded or pressed against a cation-exchange membrane [234]–[236]. The assembly formed by the membrane and the porous catalytic sheet is in turn bonded to or pressed against a suitable porous current collector, such as an expanded metal or wire mesh. The dimensions of the voids in the expanded metal or wire mesh are preferably small in order to allow simultaneously good mechanical support of the membrane–porous sheet assembly and homogeneous current distribution.

The current collector is made of a material that resists the corrosive action of sulfuric acid, which may contact the collector in the event of a discontinuity in the membrane, such as a pinhole or crack, formed during fabrication, handling, or operation. Suitable materials capable of operating at 60–70 °C are nickel alloys. With optimized assemblies (DAP, De Nora Permelec, Milan, Italy), overall anodic overvoltages of 0.1–0.2 V at 3000 A/m^2 are obtained.

In the bipolar arrangement of elementary cells the following overall reaction occurs in the separation wall between the depolarized anode and the cathode:

$$H_2O \rightarrow H^+ + OH^-$$

which is the same reaction occurring in bipolar membranes for electrodialysis [237]. In particular, the overall reaction results from an anodic partial reaction:

$$H_2 - 2\ e^- \rightarrow 2\ H^+$$

which takes place in the acid environment of the ionic groups of the cation-exchange membrane and a cathodic partial reaction.

$$2\ H_2O + 2\ e^- \rightarrow H_2 + 2\ OH^-$$

which occurs in the alkaline environment of the catholyte made of caustic soda.

When the cathode is provided with an electrocatalytic coating, the overall voltage across the bipolar separation wall between two adjacent cells has the remarkably low value of 1.2–1.3 V at 3000 A/m^2, thus equal to or even lower than that of bipolar membranes at 1000 A/m^2 [238].

The outstanding performance at high current densities largely counterbalances the intrinsically higher cost of the structure comprising the hydrogen depolarized anode, bipolar wall, and activated cathode as compared with bipolar membranes. Further advantages of the assembly are the long service life due to the mechanical strength, a low sensitivity to impurities contained in the feed electrolyte, and high quality of the products.

Operating Conditions. Voltage vs. current relationships for three types of cells provided with DAP, electrocatalytic cathodes, electrolytic compartments having 3-mm gaps, and turbulence promoters are given below:

Voltage (V)–current density (CD) relationships for three internal structures of the elementary cell for sodium sulfate electrolysis:

1) Two-compartment, cation-exchange membrane $V = 0.96 + 0.59 \times CD$ (kA/m^2)
2) Three-compartment, cation- and anionexchange membranes $V = 0.97 + 1.02 \times CD$ (kA/m^2)
3) Three-compartment, two cation-exchange membranes (buffer compartment) $V = 0.98 + 0.89 \times CD$ (kA/m^2)

The three relationships apply to current densities up to 3000 A/m^2. The DAP assembly comprises a hydrogen depolarized anode and a Nafion 117 membrane; the electrolytic compartments are limited by Nafion 324 and Selemion AAV ionexchange membranes. The relationships further apply to operating temperatures of 60–70 °C and to 13–18 % caustic soda, 200–300 g/L sodium sulfate, and 10 % sulfuric acid.

The current efficiency for the production of caustic soda ("cathodic efficiency") is a complex function of the following factors:

1) *Caustic soda concentration.* Other conditions being the same, the cathodic current efficiency decreases as the caustic soda concentration increases to more than 15–18 %.
2) *Molar ratio (MR) between sulfuric acid and sodium sulfate at constant sodium sulfate concentration.* The second dissociation constant of sulfuric acid is rather low, in the range of 0.01. Consequently, in a solution containing both sulfuric acid and sodium sulfate at MR < 1, substantially all the sulfuric acid reacts with the stoichiometric amount of sodium sulfate to give sodium bisulfate (buffer action). Hence, the actual concentration of free protons (H$^+$) is directly proportional to the actual concentration of sodium bisulfate and inversely proportional to that of the unreacted sodium sulfate. This type of dependence indicates that the actual concentration of free protons should increase quickly when MR exceeds a certain critical value (ca. 0.5). At higher MR values the current transported by the protons becomes significant at the expense of that transported by the sodium ions, and the cathodic efficiency shows a sharp decrease.
3) *Sodium sulfate concentration at constant MR.* At MR values < 1, the proton concentration, which is a function of the ratio MR to (1 – MR), is expected to remain constant at constant molar ratios, even if the concentrations of both sulfuric acid and sodium sulfate are greatly changed. Consequently, the transport number of the protons decreases as the concentration of sodium sulfate (i.e., the concentration of sodium ions) is increased. At the same time, the cathodic current efficiency increases.

This behavior has also been discussed in terms of alkaline state (to the left of the critical point of MR) and acid state (to the right of the critical point) of the cation-exchange membrane [232], [239]. This difference in internal states of the membrane is connected to the presence in the conductive channels of the membrane of a population of either hydroxyl ions (back migration from the catholyte when the electric current is substantially transported by sodium ions; low MR values) or protons (significant portion of the current transported by protons; high MR values). The model has remarkable practical consequences, particularly in terms of the sensitivity of the membrane with respect to impurities contained in the feed solutions.

The above considerations also apply to other salts that derive from weak acids, such as phosphates, acetates, and in general the sodium salts of many organic acids. The behavior of salts deriving from strong acids, such as nitrates, chlorates, and perchlorates, is substantially different because these salts are not able to perform any buffer action. Thus, at a given cathodic current efficiency, the allowed concentration of acid is significantly lower than that typical of sodium sulfate solutions. The concentration of acid can, however, be increased by the addition of a "dead load" of sodium sulfate, which creates the necessary buffer action [240].

With two-compartment cells the cathodic efficiency also represents the total current efficiency of the process. In the three-compartment cells, the current efficiency for sulfuric acid production (anodic efficiency) must also be taken into consideration. The overall current efficiency of electrolysis is represented by the lower of the two efficiencies. If, for example, the anodic efficiency is lower than the cathodic, acidity builds up in the sodium sulfate compartment during operation so that the cathodic current efficiency progressively decreases; when it reaches the same value as the anodic, stationary process conditions are maintained. Therefore, the process is provided with a self-regulating mechanism. Alternatively, if the accumulation of acidity in the sodium sulfate compartment cannot be allowed, then a portion of the caustic soda produced can be fed into the sodium sulfate loop. The quantity of caustic soda available for use outside the electrolysis plant again represents the overall current efficiency, which still remains a function of the anodic efficiency. The anodic and cathodic current efficiencies can be made independent when the control of acidity in the sodium sulfate compartment is performed by means of sodium carbonate addition.

Auxiliary Sections. The electrolysis plant may contain auxiliary units necessary for functional integration with the upstream plants where sodium sulfate is generated and the electrolysis products are fed back.

Treatment of Sodium Sulfate Feed. Impurities normally present in the sodium sulfate feed are Ca, Mg, Al, and Si. These impurities penetrate into the cation-exchange membranes separating the sodium sulfate and the caustic soda compartments; they react with the internal alkaline environment and precipitate, causing a voltage increase and a loss of cathodic efficiency.

Internal precipitation is less likely when the caustic soda contained in the cathodic compartment is diluted. With caustic soda concentrations in the range of 10–15 %, the total concentration of impurities may reach a maximum value in the order of 1–5 µg/g, depending on the type of membrane [228]. If the membrane is operated in the acid state [232], [239], this limit of concentration may be increased to 20–30 µg/g [232], [241]. Obviously, this is a compromise, because the advantage of operating under safe conditions for the membrane (even in the presence of relatively high impurity concentrations) is counterbalanced by the lower cathodic current efficiency typical of the acid state of the cationexchange membranes.

These general considerations must, in any case, take into account both the kind of electrolysis and the structure of the cells (two or three electrolyte compartments). In the first case, the sulfate solution is strongly acid and high levels of impurities are allowable. Besides, the electrolysis cell is inserted in an open cycle where the accumulation of impurities is negligible.

In the second case, the cell is inserted in a substantially closed loop. This situation leads to the accumulation of impurities. In this case, a chemical purification unit must be introduced, which, in its simplest form, may be quite similar to the primary treatment of chlor–alkali plants. The treatment is based on additions of caustic soda and sodium carbonate followed by filtration, and it allows the concentration of impurities, such as calcium and magnesium, to be kept in the range of 1–5 µg/g. The additional investment required for chemical treatment is rather limited as are the operational costs.

Chemical treatment may be replaced by crystallization or ion-exchange units. However, their investment and operating costs are substantial and negatively influence the economics of the entire process. In addition, crystallization and ion exchange are delicate operations that may require specialized personnel.

Concentration of Sodium Sulfate Solutions. A concentration unit is always necessary for both two- and three-compartment cells when the feed consists of dilute sodium sulfate solutions, because the cell voltage increases rapidly when the sodium sulfate concentration is <10%. The methods available are reverse osmosis, electrodialysis, and evaporation. With the present average European prices of steam and electric energy, evaporation seems to be the best method for concentrating up to 20% diluted feed solutions having a salt content around 3–5%.

Even if the feed solutions are concentrated (e.g., to ca. 20%), evaporation is still required when electrolysis is carried out in three-compartment cells. In this case, evaporation is directed to keep constant the amount of water in the electrolysis section. Water is let in with the feed solution and only partially migrates through the cationic and anionic membranes to form caustic soda and sulfuric acid.

Obviously, evaporation is not required when the makeup consists of solid salts. The evaporation unit is expensive with respect to both investment and operating costs, and its installation should be the object of careful economic analysis.

As a purely indicative example, an electrolysis plant based on three-compartment cells is considered, having a capacity of 3000 t/a of sodium sulfate (100% basis) fed to the cells as a 20% solution: The investment for a doubleeffect evaporation unit represents about 10–15% of the total investment for the complete plant, comprising cells, rectifier, pretreatment unit, evaporation, and construction. The cost of steam amounts to about 50% of the total electric energy cost.

Clearly, the energy consumption, which is already substantial for the 20% feed solution, increases sharply as the feed solution becomes more dilute.

Before beginning an economic analysis to define the most convenient trade-off between evaporation and electrolysis cost, the upstream plants where sodium sulfate is generated should be inspected carefully to determine whether the dilute solutions are formed as such or are the result of mixing with other process wastewaters. Internal dilution is often a consequence of the old attitude toward by-product salt solutions, which were diluted as much as possible within the battery limits of the plants to reach the limit allowed for discharge.

Sulfuric Acid Concentration. In three-compartment electrolysis, 10 to 15 wt% of pure sulfuric acid is produced with an anodic efficiency in the range of 60–75%. Often these solutions are too dilute to be fed directly to upstream plants. This requires concentration by evaporation, which, however, may be limited to a maximum concentration of 75 wt %. A unit of this type with an annual capacity of ca. 2000 t of acid (100 % basis) involves an investment that amounts to about 15–20 % of the total investment required for the complete plant. The steam consumption for a double-effect unit, at current European prices for electric energy and steam, corresponds to about 30 % of the electrical energy cost of electrolysis.

Conclusions. The Hydrina technology developed by De Nora Permelec for low-cost splitting of neutral salts has the following advantages:

1) Easy integration of the electrolysis process into the production cycle, thereby forming a "closed loop"
2) Transformation of by-products into valuable chemicals such as caustic soda and acid
3) Elimination of any environmental problems deriving from the disposal of by-product salts

3.2.13. Ultrapure Isopropanol Purification and Recycling System (Example from Mitsubishi Chemical)

Background of Isopropanol Recycling System. In the electronics industry, ultrafine pattern processing technology for greater-capacity devices has been developed recently. This technology requires complete exclusion of impurities from the manufacturing lines, i.e., a highly sophisticated clean room technique, and highly purified water and solvents among other measures.

One of the solvents used in ultrafine pattern processing technology is ultrapure isopropanol (IPA), which is widely used in vapor dryers for drying the electronic devices and liquid crystal display substrates that have been washed with ultrapure water [242]–[244]. At present, the large amount of used IPA—containing impurities such as water, metal ions, and particles—is discarded after use. When using IPA, the following serious problems arise:

1) Increase in production costs through the use of large quantities of ultrapure IPA that are not recycled
2) Latent danger of using a large amount of flammable liquid
3) Environmental problems because of the large quantities of waste IPA that are disposed of

To solve these problems, the new IPA purification and recycling system has been developed, which can reproduce ultrapure IPA that meets the specifications of the electronics industry from the waste IPA. This system is composed of a high-performance dehydration unit based on pervaporation membrane separation; a metal-ion elimination unit based on distillation; and a particle elimination unit based on micro-

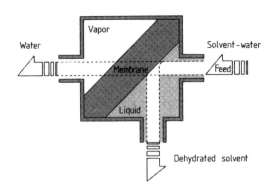

Figure 106. Flow sheet of the pervaporation process

filtration. These units remove each impuritiy efficiently from waste IPA, and purified IPA can be recycled to manufacturing processes of the electronics industry.

Pervaporation Membrane Separation. Pervaporation (PV) is a membrane process to separate the components of liquid mixtures; a liquid mixture is fed to one side of a membrane, permeating as vapor to the opposite side to which a vacuum or sweeping gas is typically applied (see Fig. 106). This nonequilibrium process is regarded as a suitable alternative process for the separation of organic liquid mixtures such as azeotropic mixtures, mixtures having close boiling points, mixtures of structural isomers, and of heat-sensitive organic compounds. Pervaporation has been commercialized for the separation of water from concentrated alcohol solution, primarily ethanol and IPA. Other applications, such as separations of small amounts of organic solvents from contaminated waters and separations of purely organic mixtures, are under development.

The most widely accepted model for describing the transport mechanism through the pervaporation membrane is a modified solution–diffusion model, where a permeating component is first dissolved or adsorbed in the membrane and then transported through it by a diffusion process. In this process, separation of the mixture is achieved by differences in the solubility and diffusibility of the individual components.

Performances of PV membranes are represented by parameters such as separation factor, flux of permeates, and service life. The separation factor of a membrane is a measure of its permeation selectivity (permselectivity) and is defined as the ratio of the concentration of components in the permeate mixture to that in the feed mixture. The component flux is the amount of a component permeating per unit time and unit area, and is given by the product of the permeability coefficient of the membrane and the driving force. The driving force is the gradient in the chemical potential of the components between the feed and the permeate side of the membrane. These values are influenced by operating variables such as temperature, composition of each component in the feed mixture, and permeate side pressures (see Fig. 107).

Separation of IPA and Water by Pervaporation. In the PV membrane separation unit, the water permselective membrane is used for dehydration. Water in the feed

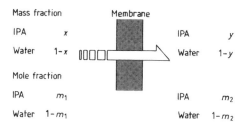

Figure 107. Mass-transfer model for pervaporation of an IPA–water mixture
J_W = Flux of water; J_i = Flux of IPA; Q_W = Permeation coefficient of water; γ_W = Activity coefficient of water; P_{0W} = Saturated vapor pressure of water at operating temperature; P_{1W} = Partial pressure of water on feed side; P_{2W} = Partial pressure of water on permeate side; P_2 = Total pressure on permeate side

Separation factor	$\dfrac{\frac{1-y}{y}}{\frac{1-x}{x}} = \dfrac{J_W}{J_i} \quad \dfrac{1-x}{x} = \dfrac{18(1-m_1)}{60\,m_1}$	
Flux of water	$J_W = Q_W \times (P_{1W} - P_{2W})$	gm^{-2} h^{-1}
Partial pressure of water on feed side	$P_{1W} = (1 - m_1) \times \gamma_W \times P_{0W}$	Pa
Partial pressure of water on permeate side	$P_{2W} = (1 - m_2) \times P_2$	Pa

mixture can permeate through the membrane; the IPA, in contrast, effectively cannot. Consequently, the residue IPA on the feed side is dehydrated.

The water concentration of the feed side decreases as dehydration progresses and lowers the chemical potential of water on the feed side. This means that a greater degree of dehydration requires a higher operating temperature to increase the driving force. The low heat resistance of conventional membranes has up to now prevented PV separation processes at high temperature [245], [246]. In contrast, the novel heat-resistant membrane used in the IPA–water system enables the operation to be performed at higher temperature, achieving not only the high degree of dehydration but also the large water flux and increasing total system performance (see Fig. 108).

The permeate water vapor is continuously condensed in a cooler, stored in a tank, and discharged. Due to the superior permselectivity of the membrane, the amount of IPA in the permeate mixture becomes very small; therefore, the IPA load of the wastewater decreases significantly.

The membrane is in the form of a hollow fiber (see Fig. 109), which has the advantage of reduced outer dimensions together with a large membrane area. The membrane module consists only of the hollow fiber bundle and the module housing. Such a simple structure can avoid difficulties encountered with other membrane module designs, such as sealing of flat seat type and spiral-wound type membranes and furthermore can reduce not only the volume and the weight of the modules, but also the total system size.

Distillation Unit and Filtration Unit. In this system, purification and recycling of IPA is performed by the combination of the PV membrane dehydration unit, the distillation unit, and the microfiltration unit. In the distillation unit, impurities, which are difficult to separate in the PV membrane separation unit such as dissolved metal ions and high-boiling impurities, are completely eliminated.

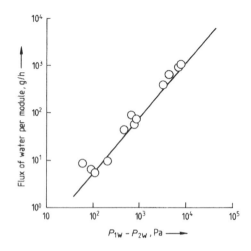

Figure 108. Relationship between difference in partial pressure of water and flux of water

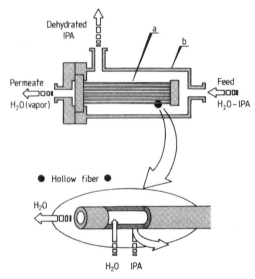

Figure 109. Structure of the pervaporation module
a) Fiber bundle; b) Module housing

In the last process step, fine particles are removed by the microfiltration unit. In the manufacture of highly integrated electronic devices, particles from the solvents used in these processes must be removed to improve product yields and suppress wafer contamination defects. For example, particles with >0.05-μm diameter should be removed to the extent of less than 10 particles per milliliter from solvents used in 16-Mbit level production lines [247]. Accordingly, the level of the microfiltration unit affects total system performance; therefore, the unit should be equipped with an appropriate filtration membrane, although only a few membranes with sufficient performance are available [248]. In the solvent a very low level of dissolved metals and low total organic carbon (TOC) is desired. Moreover, high chemical resistance of the filtration membrane is also needed.

The system can vary the required capacity by a change in its composition. For example, if the 90 % IPA makeup solution could be purified up to 99.9 % IPA concentration with a feed rate of 12 kg/h, then more than 10 kg/h of product would be obtained. The number of particles with diameter of > 0.2 μm is less than 10 particles per milliliter product. The metal ion content of the product is ≤ 1 ppb (i.e., 1 μL/m^3) for the following elements:

Al	Ca	Cu	Mg	Pb
As	Cd	Fe	Mn	Sn
B	Co	K	Na	Sr
Ba	Cr	Li	Ni	Zn

These results were achieved by using S 111 and M 111 type purification/recycling systems; the dimensions (mm) of the systems are 600, 2300 width, 1700 height and 800 depth, 2300 width, 1460 height respectively, and the masses of both systems are 500 kg. A schematic of the entire unit is shown in Figure 110.

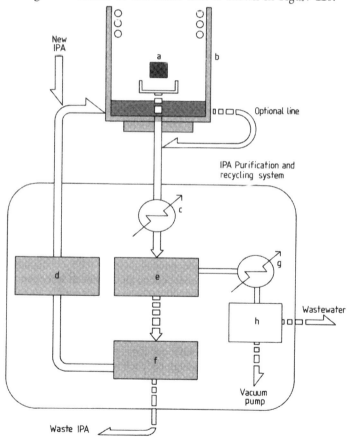

Figure 110. Typical application of the IPA purification and recycling system
a) Electronic device to be dried; b) Vapor dryer; c) Heater; d) Microfiltration unit; e) Pervaporation unit; f) Evaporation unit; g) Condenser; h) Tank

3.2.14. Examples from Boehringer Mannheim

3.2.14.1. Biocatalytic Splitting of Penicillin [249]

Introduction. The technique of biocatalysis has been used for thousands of years in the production of various food products including bread ("sourdough"); the fermentation of sugar from fruit and fruit pulp to produce alcoholic beverages; the processing of milk to produce cheese, yoghurt, and other products; the production of vinegar and sauerkraut; and many other products.

These processes utilize the activity of biological catalysts (enzymes), which are formed by bacteria, yeasts, and molds. Each individual enzyme performs a different biocatalytic function.

A particularly successful and technically elegant biocatalytic process is the conversion of natural penicillin into a product that is necessary for the production of chemically modified penicillins. This product, known as 6-aminopenicillanic acid, is formed by splitting off the side chain of penicillin.

6-Aminopenicillanic acid is treated to obtain the so-called semisynthetic penicillins, of which the most well known are ampicillin and amoxycillin (see below).

Penicillin G → 6-Aminopenicillanic acid → Ampicillin, Amoxycillin

The enormous importance of penicillins in combating infections is shown by the fact that > 4000 t/a of 6-aminopenicillanic acid alone is produced.

Today, by use of genetic engineering, the active enzymes can be obtained relatively simply and selectively in pure form. By immobilizing these in a porous matrix, they can be converted into a form with very convenient properties for technical handling purposes.

In the biocatalytic process for splitting penicillin described here, genetically engineered penicillin amidase is used, which is bonded to polymer particles the size of sand grains.

This process has almost completely superseded an older chemical process, over which it has a number of advantages including a better environmental balance:

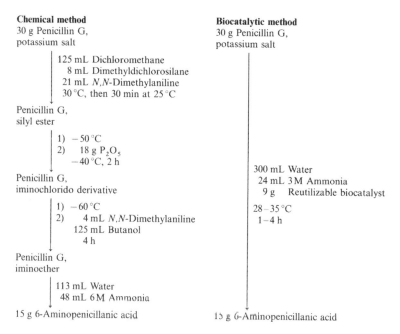

Figure 111. Reaction scheme for chemical and biocatalytic splitting of penicillin

1) Lower production costs
2) Energy savings
3) Fewer safety problems
4) Fewer disposal problems
5) Improved product quality

Comparison of the Environmental Balance of the Chemical and Bioctalytic Processes. The reason for the *reduction in production costs* is immediately apparent from the reaction sequence shown in Figure 111. Whereas the chemical process requires three intermediate stages, the biocatalytic process occurs in a single synthesis step in water at room temperature with ammonia addition for neutralization.

The advantages of the biocatalytic process are impressive if the conversion of 1000 t penicillin is considered; this yields ca. 500 t product or slightly less than one-eighth of annual world production (see below).

Chemical method	Biocatalytic method
Consumption of reagents	
1000 t Penicillin G, potassium salt	1000 t Penicillin G, potassium salt
800 t N,N-Dimethylaniline	
600 t Phosphorus pentachloride	
300 t Dimethyldichlorosilane	
	45 t Ammonia
	0.5 – 1 t Biocatalyst
Consumption of solvents (recoverable, partly for disposal)	
4200 m³ Dichloromethane	10 000 m³ Water
4200 m³ Butanol	

Product isolation and purifiction	
Acetone	Acetone
Ammonium bicarbonate	Butyl acetate
	Hydrochloric acid

Energy consumption (estimated from cooling costs)	
6×10^6 kW · h	3×10^6 kW · h

The potential *energy savings* can be estimated from the much reduced cooling costs. Whereas the chemical process requires a temperature of -50 °C, the biocatalytic process is carried out at 28–35 °C. Thus, expensive cooling is replaced by comparatively cheap heating.

The absence of highly reactive chemicals in the biocatalytic process reduces the complexity of *safety measures*, which would otherwise be considerable. Phosphorus pentachloride especially poses a problem, because it must be stored and handled with complete exclusion of moisture since it is strongly hygroscopic and decomposes with the liberation of hydrogen chloride gas.

In the biocatalytic process, waste gas purification is not necessary, and no *disposal problems* involving environmentally harmful chlorine-containing solvents or by-products from the synthesis process occur.

In the bioctalytic process, the *product quality requirements* are considerably easier to comply with. In the old process, *N,N*-dimethylaniline had to be removed from the reaction product, because this toxic substance may be present only below its toxicity threshold (i.e., in extremely small traces).

The absence of significant quantities of environmentally problematic substances and the large contribution to energy savings reflect the enormous potential of biocatalysis for solving and preventing environmental problems.

3.2.14.2. Production of Diagnostic Reagents by Means of Genetic Engineering: Glucose-6-Phosphate Dehydrogenase and α-Glucosidase [249]

Introduction. Diagnostic reagents are important tools for medical investigations. For some of these (e.g., the testing of blood and other bodily fluids), chemical reactions are no longer used; biochemical methods based on enzymes are preferred.

Glucose-6-phosphate dehydrogenase (G6PDH) (E.C. 1.1.1.49) is used in diagnostics to determine the glucose content of serum, and *α-glucosidase* is used in a test combination for the determination of amylase content, an important parameter in determining the functional capability of the human pancreas. High-purity enzymes in kilogram quantities are required for the enormous number of laboratory investigations conducted.

Conventional biochemical processes require up to 1000 t of biological raw materials to produce 1 kg of an enzyme suitable for bioanalysis.

The efficiency of enzyme production by biotechnological methods can be much improved by genetic engineering. Reprogramming (cloning) the genetic information

Table 7. Prevention of environmental pollution by use of recombinant organisms in the producton of G6PDH during fermentation

Quantity	Leuconostoc	Recombinant E. coli
Fermentation volume, m^3	600	1
Salts and nutrients, kg	64 000	160
Pure water consumption, m^3	120	1
Cooling water consumption, m^3	15 000	30
Amount of O$_2$-free water required, m^3	740	
Amount of ice water required, m^3	4 800	
Energy requirements for steam production, kW · h	16 000	270
Compressed air, m^3	114 000	570
Wastewater PE* value	300 000	300

* PE = population equivalent (i.e., organic load due to a member of the population over 24 h).

of microorganisms that can be grown by simple means (bacteria or yeasts) enables them to produce certain proteins (e.g., enzymes) in excess. By means of recombinant DNA technology, up to 1000 copies per cell can encode the desired protein. In this way, up to 50 % of the total protein in the cell can consist of the desired product. By culturing the genetically engineered microorganisms under special conditions in a medium containing special nutrients, the overproduction of the cell can be utilized technologically. At the same time, large savings in energy and raw materials are made.

Reduction of Environmental Pollution by Production of Glucose-6-phosphate Dehydrogenase and α-Glucosidase by Genetic Engineering. *Production of Glucose-6-Phosphate Dehydrogenase.* The bacterial strain *Leuconostoc* sp., used in the production of G6PDH by the conventional biochemical method, has a volume activity of the desired enzyme of 5.5 U/mL. By cloning the appropriate gene in *Escherichia coli*, an increase in the genetic enzyme yield (genetic expression) to 500 U/mL is obtained even on the 5-mL scale. Optimization of the fermentation process enables a further increase by a factor of 1000 to be achieved, i.e., a genetic expression of 4500 – 5000. The savings in raw materials and energy during fermentation alone resulting from using *E. coli* instead of *Leuconostoc* are listed in Table 7.

A comparison of the conventional culture process with that using recombinant *E. coli* shows that only 0.1 % of the raw materials and energy of the conventional process is required for the recombinant process. This is true for both the fermentation process and the downstreaming procedure (Table 8).

To provide a given quantity of this enzyme, only 200 kg biomass is required when using recombinant *E. coli* instead of 22 000 kg. Savings of a similar order of magnitude are obtained in the areas of energy consumption, steam, water, pure water, and of course, electricity.

The recombinant process gives a saving of ca. 95 – 98 % in raw materials and energy. This leads to a reduction in the load on the sewage treatment plant which is illustrated by so-called population equivalents. A population equivalent (PE) is defined as the organic load due to a member of the population over 24 h. Treatment of this organic load requires 60 g of oxygen. In the production of G6PDH by recombinant organisms,

Table 8. Prevention of environmental pollution by use of recombinant organisms in the production of G6PDH during workup

Quantity	Leuconostoc	Recombinant E. coli
Biomass, kg	22 000	200
Wastewater, m^3	1 200	0.2
Water, m^3	300	10
Ice water, m^3	4 000	50
Pure water, m^3	300	10
Steam, t	180	10
Electricity, kW·h	4 000	100
Ammonium sulfate, kg	13 000	200

Table 9. Prevention of environmental pollution by use of recombinant organisms in the production of α-glucosidase (1988 consumption figures)

Quantity	Production of α-glucosidase	
	From new yeast	From recombinant yeast
Yeast consumption, t	236	10
Wastewater from yeast production plant, t (ca. 9 t wastewater/t yeast)	ca. 2 000	ca. 90
Quantity of yeast residues for disposal, t	440	12
Ammonium sulfate, t	1 100	25
Potassium phosphate, t	25	0.5
Aluminum gel, t	90	
Filter aids, t	133	5
Solid waste, t	540	18
Liquid waste, t (ammonium sulfate + potassium phosphate as fertilizers)	1 125	25.5
Deionized water, m^3	3 700	50
Ice water, m^3	52 000	2 000
Electric power, kW	44 500	9 000
Steam, t	220	50

the organic load produced by the fermentation process is equivalent to 0.1% of that produced by the conventional process. This means that the amount of organic waste to be treated, instead of being equivalent to that from a large town (300 000 PE), is only equivalent to that from several large families (300 PE).

Production of α-Glucosidase. α-Glucosidase can also be produced very effectively using recombinant yeast instead of normal yeast strains (Table 9).

The treatment of residues from processes based on genetic engineering (wastewater, decomposed biomass) must be carried out in accordance with the safety requirements of genetic engineering law and genetic engineering safety regulations. The genetically modified microorganisms (*E. coli* and yeast) used in the processes described here are classified in the lowest safety group S 1. The wastewater (fermentation liquors) can therefore be transferred directly to a biological sewage treatment plant. Solid residues (decomposed biomass) can be composted without further treatment.

When using recombinant yeast strains, the amount of biomass to be disposed of is much smaller than with normal yeast strains because the fermentation volume is smaller (i.e., 1.2 t instead of 440 t). The water and energy consumption in the α-glucosidase production process involving genetic engineering is ca. 10–20% of that of the conventional process (Table 9).

Today, about 100 diagnostic products are produced by Boehringer Mannheim in this environmentally friendly way using genetic engineering techniques.

4. Waste Management in the Chemical Industry

4.1. Introduction

The central function of the chemical industry is the economical transformation of substances into products desired in the market. This transformation should take place with a yield as close to 100 % as possible, but there is always a percentage of undesired products, which are collected under the term "residues."

A second important task of the chemical industry is to minimize the impacts of residues on the environment as far as is technically feasible. To carry out this task, the industry employs waste-management concepts which encompass waste avoidance by avoiding or reducing the quantity of residues and the recycling of residues, as well as the disposal of the remaining residues as "waste." Avoidance, reduction, and reclamation take priority over disposal by incineration and landfilling. A final task of waste management is avoiding or reducing the quantity of production wastes and disposing of them safely.

In the downstream processing, conversion, and consumption sectors, some product becomes waste, and this is referred to as "product wastes" to avoid confusion with production wastes. Recycling of product wastes is acquiring greater importance and is becoming a further waste-management function in the chemical industry.

4.2. Chemical Industry Wastes

The industrial production sector in Germany generated 217.8×10^6 t of waste in 1993 [250]. This included 6.03×10^6 t of waste requiring special monitoring, which is also referred to as hazardous waste. In the same period, the German chemical industry, as part of the production sector, generated 5.08×10^6 t of waste [250], including 1.64×10^6 t of hazardous waste. The broad production spectrum of the industry means that the wastes are also very diverse in terms of chemical composition and consistency. This is illustrated by the following list of the most important wastes (1993, in 10^3 t) [251]:

Sewage sludges, sludges from wastewater and water treatment (dry basis)	883
Building rubble, excavated soil, pavement rubble without harmful contaminants	903
Wastes from energy production (slags, ashes, filter dusts)	141
Red mud	563
Residential-type commercial waste (incl. used paper, packaging, industrial sweepings)	280
Building rubble, excavated soil, soils with harmful contaminants	305
Inorganic acids, acid mixtures	6

Gypsum/lime mud with harmful contaminants	153
Sludges from precipitation and solution processes with contaminants	75
Organic solvents and solvent mixtures	
containing halogenated solvents	81
not containing halogenated solvents	171
Filter aids and absorbents with contaminants (kieselguhr, activated carbon, activated earths)	12
Organic distillation residues containing solvents	133
Carbide sludge with no harmful contaminants	3
Barium sulfate sludge and rock-salt residues (gangue)	49
Plastics wastes	39
Packaging material with harmful contaminants	17
Wastes from plastics manufacturing and converting	13
Lime and gypsum mud with no harmful contaminants	16
Oil and gasoline separator contents	9
Organic dyes	7
Plastics slurries	94
Aliphatic amines	11
Filter dusts containing nonferrous metals	2
Sludge from tank cleaning	5
Other wastes	1113
Total	*5084*

Over half the quantity of wastes is accounted for by three large groups: 27% wastes not associated with chemical production (building rubble, excavated soil, wastes from energy production facilities); 17% from wastewater and water treatment; and 16% inorganic sludges.

This quantitative information represents a waste balance which deviates from the waste statistics, and is defined as follows [250]:

1) Duplication is removed by subtracting wastes accepted by other businesses, and secondary wastes generated in waste treatment (e.g., slag and ash from incineration)
2) Quantities delivered by waste producers to reprocessors or scrap dealers (i.e., residues) are not included
3) Sewage and water-treatment sludges are reported on a dry basis because of their high water content

Comparable statistics broken down in a similar way are not available for the chemical industries of other countries.

4.3. Waste-Management Concepts

4.3.1. Residues and Wastes from Production

A chemical production process generates residues in addition to the intended end product. These residues can be recycled (Fig. 112). If recycling is impossible for technical or economic reasons, residues become waste. In addition, polluted waste-

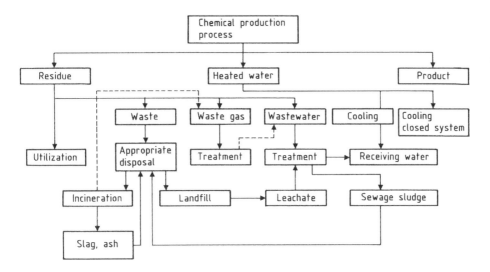

Figure 112. Principles of chemical production processes

water and waste gases can be formed. Disposal of wastes includes incineration and landfill disposal. Waste incineration generates secondary wastes in the form of slag and ashes, which have to be utilized or disposed of. Wastewater is produced in waste-gas scrubbing. The treatment of wastewater leads to the formation of sewage sludge and other sludges, which must then be disposed of as waste. Landfill leachate must also be treated, and once again this process can contribute to sludge formation. In this way, a complex residue/waste system results, and the solution of one problem results in the formation of secondary waste [94].

4.3.2. Production-Oriented Management

Residues and wastes can arise in more than one environmental area of production. Concepts for residue management must therefore extend over many such areas. Disposal of residues as wastes, considered alone, even if carried out in an appropriate, environmentally compatible form, has the disadvantage that the waste problem is dealt with at the end of the production chain; there may also be ecological and economic drawbacks associated with this arrangement. These end-of-pipe technologies are generally more expensive than upstream measures.

Waste management in production is a planned, purposeful technique including three elements [252]:

1) Analysis of wastes and waste production by type and quantity in the production chain. This makes it possible to device an "avoidance" concept.
2) Environmentally compatible technical measures (disposal) or new solutions for the waste problem on the basis of the analyses.
3) Monitoring of the practices implemented.

Today, the principle of waste avoidance is incorporated in new process development (efforts to avoid generating residues or decrease the quantity generated). Existing processes are also studied and reconfigured if appropriate. This is the principle of production-integrated environmental protection. Another way to implement this principle is to utilize residues; the relative priority of utilization and waste disposal has to be established case by case. The ecological and economic effects of utilization have to be analyzed; one question is whether reclamation of materials may harm the environment more than it helps, for example when recovery processes consume disproportionate amounts of energy. Again, many reclamation processes generate wastes, wastewaters, and waste gases, which in turn must be disposed of. Finally, a reclamation process cannot be repeated indefinitely, and utilization has to be viewed from the market standpoint. Accordingly, it is necessary to clarify in advance whether the reclaimable substance is required and accepted by downstream processors or consumers. Problems can arise when the new material obtained by reclamation is of lower quality than the primary product. If there is no market, even the best reclamation concept is doomed to fail. Even a viable reclamation concept can come to nothing if the level of acceptance changes abruptly. The decision between utilization and disposal must also take account of ecological balances prepared by scientifically sound and objective methods. This statement also holds for thermal utilization, which ranks equally with recycling of substances: A residue is thermally utilizable if its combustion produces more heat than the sum of the losses experienced. The difference between the heating value of a substance and the unavoidable losses in combustion can be recycled to the production process in the form of steam and electric power. This energy replaces primary energy sources such as coal, petroleum, and natural gas. If the energy balance is zero or negative, however, thermal treatment becomes a form of disposal.

Because residue reduction and reclamation cannot lower the quantity of production waste to zero, environmentally sound disposal will remain the basis of waste management for the foreseeable future. Disposal takes the form of incineration or landfilling. Each of these activities today meets high technological and safety standards and ecological concerns are amply addressed. Emissions into air, water, and soil have been cut to a very low level. A modern disposal concept also includes analysis of the present quantity, makeup, and properties of the waste and waste streams along with short-term projections. Whenever possible, secondary wastes such as ash and slag produced by incineration are also utilized, and the heat produced is reclaimed in the context of energy recycling.

4.3.3. Product-Oriented Management

When products of the chemical industry are processed and consumed, they are partly converted to waste. A reduction of such product wastes is necessary and can be achieved through recycling of product wastes or the formation of closed loops [253]. The chain of responsibility from producer via downstream processor and consumer must also be completed under a "holistic marketing" concept [254]. This means considering what happens to the material after use. The technical reclamation measures have to be matched to each type of product waste.

Recycling can involve recovery of raw materials by chemical cleavage or reclamation of materials with no alteration of chemical structure. It is necessary to verify that necessary cleaning, transport, and energy do not lead to ecological impacts elsewhere. Reclaimed product wastes should always fill gaps in the market and help to save costly raw materials; they should not be justified solely by recycling for its own sake. Again, market conditions must not be left out of account, and the technical constraints on the reclamation of product wastes must be taken into consideration. Plastics wastes, for example, are not indefinitely recyclable without change in composition, because the recycling processes alter the molecular structure. The structural changes in turn affect macroscopic properties of the plastic, such as toughness and strength. The recycling system takes the form of a cascade ("downcycling") that ends in the irreversible transformation of matter to energy by combustion with utilization of the energy, or disposal in landfills.

4.4. Disposal Measures

4.4.1. Logistics

Analytical studies are used to monitor and control production processes in the chemical industry and to characterize chemical products for their later use. The same holds for wastes and waste-disposal processes.

Any residue that is not utilized in the production plant is characterized by an extensive list of chemical and safety data. This information is used in assessing the behavior of the waste during shipping, storage, and disposal, and also serves as a guide for selecting the appropriate disposal method. Studies of the physical and chemical properties of wastes are performed and their exact compositions are determined. Data on the production process that gives rise to the wastes are also integrated into the process. A product-by-product breakdown of waste quantities is obtained on the basis of a production registry. In this way, when production changes are planned, the resulting changes in waste streams can be predicted as well.

When wastes are to be disposed of in landfills, a variety of parameters are ascertained by analysis of the original substance and analysis of the aqueous eluate. Other studies are related to landfill emissions such as leachate.

Some special parameters (e.g., flash point, heating value, water content) must be determined for wastes that are to be disposed of or thermally utilized by combustion. All data for a given waste are documented in a multi-page survey form. The plant manager is responsible for the correctness of the information. After thorough checking, including a check for completeness of analysis data, the necessary pretreatment and the disposal route (including the proper transportation method) are decided on by a cental corporate office. Disposal and transformation are only carried out when official approval has been obtained for each individual case. Computer-based systems are used to document all analytical data on wastes, calculate disposal costs, and provide data to the regulatory agencies.

4.4.2. Waste Combustion

Hazardous Wastes. Incinerators are used for solid, pastelike, and liquid wastes that cannot be placed in landfills because of environmentally relevant constituents, provided they can burn without exploding. Such wastes are essentially organic production residues, vegetable and animal fat products, herbicides and pesticides, petrochemical and coal products, organic solvents, dyes, paints, adhesives, putties, and resins. To avoid harmful emissions when these products are combusted, the flue gases must be cleaned.

Before initial delivery, each waste must be examined as to its relevant physical and chemical data. The purpose of this analysis is to make certain that the emission limits as well as occupational health and safety and fire-protection standards are not violated upon combustion. The results of this investigation make it possible to classify wastes, that is, to group them with respect to the combustion process and thus achieve optimal furnace conditions.

Widely used combustion systems include rotary kilns, fluidized-bed furnaces, and multiple-hearth furnaces. A number of wet and dry processes have been devised for cleaning the flue gases. Hot flue gases are cooled and their energy is recovered and used to raise steam before the cleaning stage.

At BASF, flammable solid, pastelike, and liquid residues are combusted in eight furnaces. Each combustion unit consists of a rotary kiln with afterburner chamber and a steam boiler. The superheated 18-bar steam from units 1 to 6 is fed into the BASF plant network. In units 7 and 8, a higher-value steam is generated with an efficiency of ca. 74 % and supplied to a back-pressure turbine, where it is expanded from 43 bar to 5 bar. To utilize the heat of the flue gases between 300 °C and 180 °C, a waste-heat boiler was installed to raise 5-bar steam. Electric power and 5-bar steam are fed into the respective plant systems (Fig. 113).

Three flue gas cleaning units were built for the eight rotary furnaces. Each unit consists of an electrostatic filter and a three-stage scrubber. By 1996, these units will be retrofitted with catalysts to lower NO_x emissions and degrade dioxins. The plant combusted ca. 150 000 t of waste in 1992; on reconstruction, a combustion capacity of 160 000 t/a will be available after 1996 [255]–[258].

The Hoechst Waste Incineration Plant (capacity: 46 000 t/a) is composed of two combustion facilities. The central part of both facilities is the rotary kiln. Adjoined to the rotary kiln is the post-combustion chamber in which optimal and complete combustion of the flue gas is achieved at temperatures above 900 °C. This process is

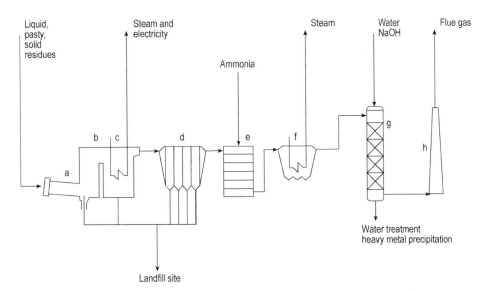

Figure 113. BASF hazardous waste incinerator
a) Rotary kiln; b) Secondary combustion chamber; c) Boiler; d) Electroststic precipitator; e) Catalyst; f) Low temperature boiler; g) Scrubber; h) Stack

controlled by precise operations of the process control systems. Recovery of energy takes place in the steam boiler.

To prevent quick ash from escaping, an electrostatic filter takes up the process and precipitates the quick ash from the flue gas.

In a two-step "flue gas wash", the gaseous inorganic pollutants, such as hydrogen chloride and sulfur dioxide, are removed and neutralized.

Subsequently, the flue gas is heated to the operating temperature of the coke filter.

The coke filter then binds the fine dust particles, the last traces of gaseous inorganic pollutant and fine traces of organic material, such as polychlorinated dioxins and furans.

The washed and filtered flue gas passes through another Selective Catalytic Reduction (SCR) catalyst, which, with ammonia, reduces nitrogen oxides to nitrogen and steam and removes the remaining traces of organic substances.

The water from the flue gas wash flows through a wastewater treatment plant in which heavy metals and dioxins are precipitated and filtered out. In this process, about 100 t of sludge arise per year. The sludge is then drained in a filter press. One percent of the remaining sludge still contains heavy metals and must be disposed of in an underground deposit.

The remaining wastewater now containing only uncritical salts, such as sodium chloride and sodium sulfate, continues through the purifying process into a biological wastewater treatment facility and is then released into the river.

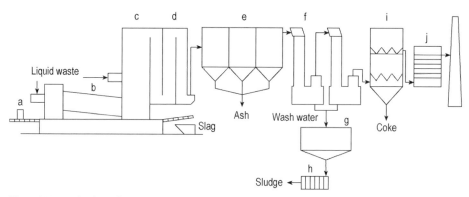

Figure 114. Hoechst hazardous waste incinerator
a) Solid and pasty residuum; b) Rotary kiln; c) Post-combustion chamber; d) Steam boiler; e) Electrostatic filter; f) Two-step flue gas wash; g) Wastewater treatment plant; h) Filter press; i) Coke filter; j) Catalyst

2 000 t of slag are quenched in the "wet slag remover" or "water bath" and are then glazed. The slag binds the inherent heavy metals and can be deposited in a special depository. 800 t of quick ash are precipitated in the electrostatic filter each year and must be deposited underground.

At present, the unpurified coke from the coke filter is deposited underground as well. However, its incineration in the rotary kiln is already sheduled for the future.

Every hour, up to 6.2 t of waste are incinerated. The combustion of one ton of waste yields, on average, 16 000 MJ. The carbon dioxide emission amounts to only half of the hourly emission rate of a commercial airplane.

Under the most unfavourable conditions, the additional pollution caused by a few kilograms of nitrogen oxides per hour is in the ppm range.

The carbon monoxide output of the two combustion facilities at full capacity amounts to approximately 2 kg/h. This output is equivalent to that of one single car without a catalytic converter.

As a result of the coke filter the concentrations of the remaining pollutants (hazardous inorganic gases and quick ash) lie beneath the maximum permissable concentrations for working areas.

The remaining amount of dioxin, would resemble the size of a pin head each year.

The heavy metal content, which is released into the river along with the water from the flue gas wash columns, has been reduced to a few kilograms per year (Fig. 114).

Bayer operates a number of similarly engineered waste combustion units. At the Dormagen plant, a further combustor for solid and liquid wastes has been placed in service. It consists of a rotary furnace, afterburner, waste-heat boiler, and gas scrubber. The plant incorporates a condensation-type electrostatic filter specially developed by Bayer [259] and an SCR unit for selective catalytic reduction of nitrogen oxides (with ammonia) and for degradation of dioxins in the tail gas (see Fig. 115).

Sewage Sludges. At BASF, ca. 100 000 t/a of sludge (dry basis) is dewatered to ca. 40 % dry content (see Fig. 116). This consists of excess sludge from the biological

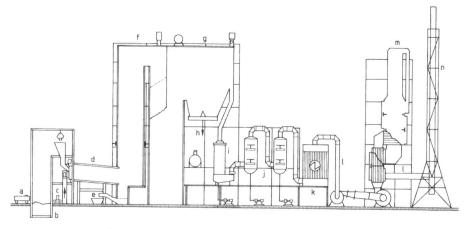

Figure 115. Bayer hazardeous waste incinerator
a) Tanks for liquid residues; b) Waste bin for solid water; c) Drums for solid waste and sludge; d) Rotary kiln; e) Slag discharge; f) Post-combustion chamber; g) Heat recovery (boiler); h) Ash discharger; i) Quench; j) Rotational scrubber; k) Condensation electrostatic precipitator (ESP); l) Exhaust vent; m) Selective catalytic reduction (SCR); n) Stack

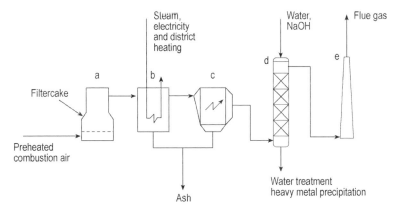

Figure 116. BASF sewage sludge incineration plant a) Fluidized bed incineration; b) Steam boiler; c) Electrostatic precipitator; d) Scrubber; e) Stack

wastewater treatment system mixed with coal and ash (filter aids). The filter cake is combusted in fluidized-bed units with a positive energy balance. The heat liberated by combustion in the furnaces is used to raise steam. Flue-gas cleaning consists of dust removal with an electrostatic filter and wet scrubbing [255], [258]. The multiple-hearth sludge combustion method used by Bayer is similar (see Fig. 117). Waste gases and vapors from the multiple-hearth furnace are thermally treated in a post-combustion chamber with addition of liquid wastes [260].

Hoechst has put a two-stage sludge combustion system in service, with a capacity of 130 000 t/a. The combustion unit is a fluidized-bed furnace consisting of the fluid-bed space proper and a separate, cylindrical post-reaction zone operated at 850 °C (see

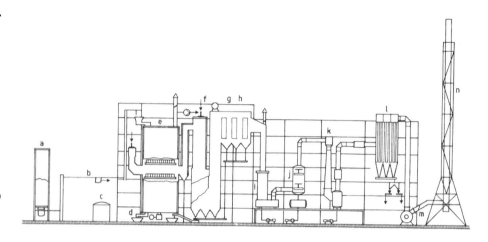

Figure 117. Bayer sewage sludge incinerator
a) Sewage sludge bin; b) Sewage sludge transport system; c) Liquid residues; d) Ash discharge; e) Multiple hearth furnace; f) Post combustion chamber; g) Flue-dust discharge; h) Heat recovery (boiler); i) Quench; j) Rotational scrubber; k) Jet scrubber; l) Adsorber bag house; m) Exhaust vent; n) Stack

Fig. 118). The oxygen content in the fluidized bed is held between 1 and 3 vol % to lower the NO_x concentration. Both carbon monoxide and organic carbon are burned in the post-reaction zone; preheated secondary air containing 6–7 vol % oxygen is injected at the head of the furnace. In this way, fuels that contain nitrogen-containing compounds (e.g., protein compounds in sewage sludge) can be burned with a much lower NO_x formation rate, eliminating the need for a flue-gas denitrification unit [261]. Flue-gas scrubbing liquor is recycled, with a substream going to a wastewater treatment unit. Heavy metals in the scrub liquor are removed there in a multi-stage process:

1) Two neutralization stages with caustic soda (precipitates sparingly soluble metal hydroxides)
2) Addition of sodium sulfide in neutralization stage (precipitates even more sparingly soluble metal sulfides)
3) Flocculation of heavy-metal precipitates with iron(III) salts and polyelectrolytes; settling and filtration

The product water is fed to the biological wastewater treatment facility.

The ashes from the fluidized bed furnace (fluidized bed ash), heat recovery boiler and electrostatic precipitator (flue ash) are collected in two separate silos. All heavy metals still present are firmly bonded in the ash.

Ashes of the sludge are used in the mining industry as an aggregate for mining mortar. This is used for backfilling worked-out mine galleries to prevent mining subsidence. Other possible uses are as a recultivation material on former spoil tips and as a construction material in mining operations.

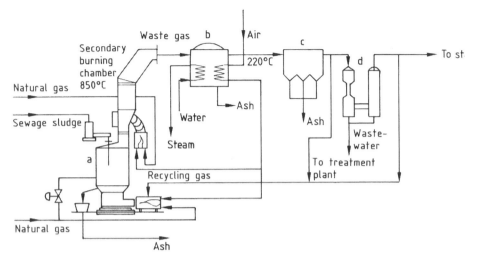

Figure 118. Hoechst sewage sludge incineration plant
a) Fluidized-bed reactor; b) Boiler; c) Electrostatic precipitator; d) Off-gas scrubber

4.4.3. Landfill Disposal of Wastes

Landfill design, the technical and organizational aspects of landfill operation, and the requirements on wastes for landfill disposal are regulated in detail by TA Abfall (an administrative regulation under the German Waste Disposal Act). The focus is protection of groundwater, which is achieved by the use of a landfill foundation seal. This feature can consist of a natural seal (e.g., a bed of clay) or an artificial liner (e.g., plastic sheeting). Leachate formed above the liner is removed through drains, collected, and forwarded to treatment. An example with an artificial seal is the Hoechst AG sewage sludge landfill (see Fig. 119). Leachate is led away via drain pipes and collected for treatment. Another landfill operated by Hoechst, for production wastes, uses a 40 m underground clay bed as its foundation seal. Leachate and precipitation are collected and sulfates are removed in a pretreatment system in which sulfate is precipitated as gypsum. The treated leachate is then forwarded to biological treatment.

For its Leverkusen landfill, Bayer devised new intermediate and cap seals. Textured plastic webs functioning as space-saving intermediate seals form an effective barrier to prevent precipitation penetrating into the wastes until the cap is placed. Each intermediate layer has its own drain system connected to a water interceptor and monitoring system. The construction of the intermediate seals began in 1989. Such seals or equivalent measures, as well as the construction of inorganic barrier layers to protect the slopes and top surface, constitute a dependable multibarrier system. All contact of the waste in the landfill with the environment is blocked. This design ensures ecologically compatible and safe deposition of wastes for the long term [262].

The BASF landfill on the island of Flotzgrün in the Rhine has a double liner of plastic webs, insuring that all leachate is collected and delivered to treatment. The

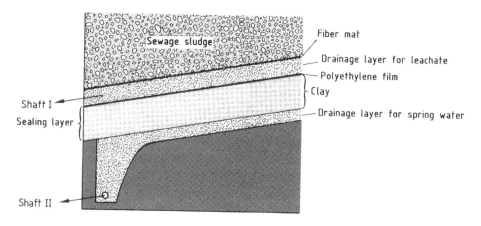

Figure 119. Landfill site with sealing layer for leachate

effectiveness of the liner is continuously monitored, and leaks can be repaired [256], [263], [264].

4.4.4. Asbestos Disposal

Solvay Umweltchemie has devised a process for disposing of asbestos (the Solvas process). Asbestos with a particle size of ≤ 5 mm is reacted with hydrofluoric acid in a batch operation, yielding fluorides and hexafluorosilicates of the elements present in the asbestos-containing material. Crocidolite (blue asbestos) and white asbestos differ in reaction time. After filtration, the solids are neutralized with milk of lime and the filtrate is recycled. Neutralization yields a product consisting of CaF_2, oxides, and metal hydroxides of the elements present in the starting material, and silicate-containing compounds. This end product can be used as an additive in the manufacture of concrete blocks and as a flux [265]–[267].

4.5. Utilization of Product Wastes

4.5.1. Plastics Recycling

Plastics to be recycled after use must be collected in the purest possible form. The recyclable portion of the waste must be separated into classes of valuable substances or individual substances. The fractions that result can be recycled for valuable constituents, recycled for feedstocks, or thermally utilized, depending on the type of plastic.

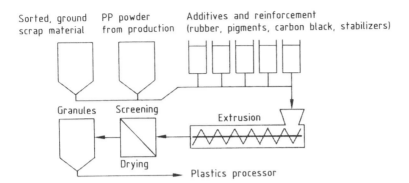

Figure 120. Recycling of polypropylene at Hoechst

Recycling of Polypropylene. At its Knapsack site, Hoechst has erected a recycling plant for polypropylene with a capacity of 5 000 t/a of PP recyclate (Fig. 120). Recycling makes it possible to start with pure PP fractions from used parts (such as those from automobiles) and modify them into "typeware." The material must be collected, ground, and sorted on a decentralized basis. The ground recycled parts are mixed with virgin PP powder, additives, and reinforcements, processed in an extruder, and thus converted to products that meet the required specifications [253].

Recycling of PVC Processing Wastes. A recycling plant for poly(vinyl chloride) wastes has been built at the Hoechst, Gendorf site. The principal feed is rigid PVC film trimmings from film processors. The plant has a capacity of 4 000 t/a. The recycling line begins with grinding of the PVC wastes and removal of foreign substances such as paper, wood, and other types of plastic. At the same time, the PVC recyclate is washed and reground. After these steps, the PVC is subjected to final grinding, formulation, and conversion to new raw material for PVC film. In the same plant, this secondary feed is converted to PVC film for nonfood packaging and technical applications. The purpose of this prototype plant is to gather experience for future plants at other locations [253].

Recycling of Polyoxymethylene. [253] The recycling techniques used up to now (regranulation and compounding) can damage polyoxymethylene. The consequences are separate specifications and application restrictions for recyclates. With a new process from Hoechst, these drawbacks can be avoided. The material recovered here is not the polymer itself but the monomer, chiefly trioxane, which is obtained by chemical degradation of POM wastes. The result of this recycling process is always new product. Only about 2 % of the POM recycling feed (consisting of stabilizers) becomes waste; this is combusted. The prerequisite for the process is that used plastic parts and processing residues be collected in pure form. A plant for the treatment of up to 1 000 t/a waste from the POM-polymerization process is in operation.

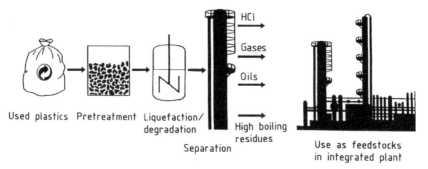

Figure 121. Recycling of used plastics into new feedstocks (BASF)

Recycling of PET Wastes. Hoechst Celanese has built a 70 000 t/a plant at Wilmington, Delaware, to recycle poly(ethylene terephthalate) wastes by methanolysis.

The glycolysis process is used in a recycling system (capacity 5 000 t/a) for PET wastes at the Offenbach plant of Hoechst.

Recycling of Used Plastics. BASF is constructing a 15 000 t/a pilot plant for recycling of mixed, unwashed used plastics into feedstocks (Fig. 121).

The conversion of mixed plastics to petrochemical feedstocks takes place in a multistage melting and degradation process. In the first conversion stage, the used plastics are melted, and dehalogenated to protect downstream equipment from corrosion. Hydrogen chloride evolved in this stage is absorbed and fed to the hydrochloric acid plant. Gaseous organic products are compressed and can be utilized as feeds to a cracker.

In the next stages, the liquefied plastics are further degraded and fed to distillation towers. The distillates include naphtha, olefins, and aromatics, which are used as feedstocks in BASF's petrochemical plant. The high-boiling bottoms are converted to synthesis gas [268]. This pilot plant was shut down, because plastic wastes in economic amounts were not available commercially.

4.5.2. Refrigerant Recycling

In 1989, Hoechst decided that its output of perhalogenated chlorofluorocarbons (CFCs) should be phased out for environmental reasons and stopped by the end of 1994. The ozone-damaging CFCs are being replaced with R134a, whose contribution to the greenhouse effect is 90 % less than that of CFCs. The new product, like the earlier ones, is integrated into a two-stage recycling system.

Recycling of used CFC refrigerants was introduced in 1986 – 87. After collection by dealers, the essential processing steps are mechanical purification and distillation.

Refrigerant mixtures that cannot be separated by distillation, together with CFC still being delivered to Hoechst after production shutdown, are subjected to secondary

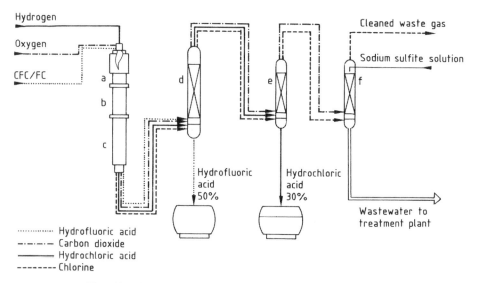

Figure 122. CFC/FC cracking plant with HCl and HF recovery (Hoechst system)
a) Combustion chamber; b) Cracker; c) Absorber; d) HF recovery; e) HCl recovery; f) Scrubber

recycling by a process that is patented worldwide. The refrigerants are cracked in a thermal reactor, and hydrochloric and hydrofluoric acids are recovered from the reactor outlet gas (Fig. 122). These acids are used as feeds for other production processes [253], [269].

4.5.3. Recycling of Used Packaging Materials

In a pilot project of the Verband der Chemischen Industrie, a facility for disposing of chemical drums was erected in the Hoechst Ruhrchemie plant. After pretreatment, drums with capacities of ≥ 120 L are cleaned, dried, and examined for damage and cleanliness. Clean drums are forwarded to recycling. Damaged drums larger than 120 (60) L as well as small containers (< 120 or 60 L) are ground, washed, and dried. The materials, in the form of pure steel scrap or sorted plastic chips, are then available for recycling [253], [270].

In 1991, BASF placed a central drum-cleaning plant in service. The next year, some 330 000 empty drums were cleaned in three automatic cleaning lines. The drums varied in size, material (plastic and steel), and design (open-head and tight-head). Two lines are run with water (with additives) and one with solvent [255].

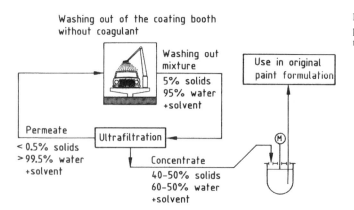

Figure 123. Recycling of paint overspray in automotive production line coating

4.5.4. Paint Recycling

Product development in water-based paints must consider how paint residues are to be recycled in the user's painting facilities. In certain paint systems, such as auto body paint booths, overspray is caught in circulating aqueous liquors and reconcentrated by ultrafiltration (see Fig. 123). Concentrate solutions are added to the appropriate paint formulations. In a special stove-drying paint system, the material recycling rate is ca. 95 %. The special membranes needed for the ultrafiltration step were developed by Hoechst subsidiaries Herberts and Hoechst Celanese Corporation, together with plant construction specialists.

4.6. Results of Waste Management

The results of waste management are characterized by two areas. The primary one relates to avoidance, reduction, and recycling of residues generated in production. This is achieved by means of production-integrated environmental protection, leading chiefly to avoidance or reduction of waste.

The second area, waste disposal, exists because no chemical plant can operate without generating waste. Techniques used here include landfill disposal when appropriate, as well as waste combustion with energy recovery and modern flue-gas cleaning. Disposal is described in terms of the type, quantity, and disposal mode of the wastes. The following information on these processes has been provided by BASF, Bayer, and Hoechst. As a result of organizational changes at Bayer and Hoechst, the data for the years following 1995 cannot be compared to the data for earlier years and have thus not been given here:

BASF. The quantity of wastes combusted with energy recovery in 1983 was 94 000 t. This quantity rose steadily in later years, reaching 158 000 t in 1995 (see Fig. 124).

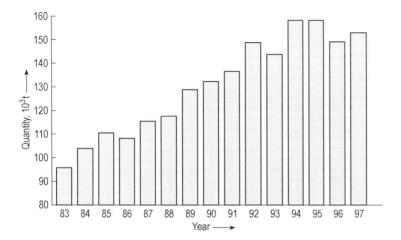

Figure 124. Waste incineration at BASF, 1983–1997

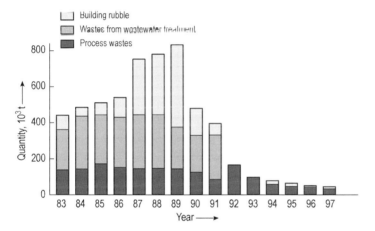

Figure 125. Quantities of wastes deposited in the Flotzgrün landfill (BASF)

Figure 125 shows the quantities of waste disposed of in the Flotzgrün landfill from 1983 to 1997. These wastes are classified as process wastes, wastewater treatment sludges, and building rubble and construction materials. The decrease in quantity since 1989 is attributed to recycling of building rubble and avoidance and recycling of process residues. Since 1992, all sludge from wastewater treatment has been combusted, so in future this source will contribute only ashes to the landfill. Other wastes deposited in the landfill in 1997 were 31 600 t of process wastes, compacted trash, trash from plant cleaning, and wastes from sludge combustion, together with 13 400 t of building rubble and construction materials [256], [271], [272].

Bayer. The total amount of wastes generated in 1995 was 695 000 t. These quantities represent a reduction from the years 1990–1995. Figure 126 gives a breakdown of

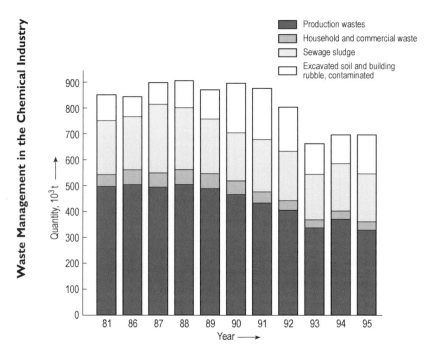

Figure 126. Waste quantities (Bayer)

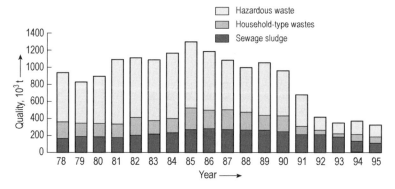

Figure 127. Waste quantities (Hoechst)

wastes with total quantities for the period from 1981 to 1995. The 7% increase in waste produced between 1981 and 1988 was essentially due to a gain in production (nearly 16% in the same period). Various practices introduced in production plants since 1988 have led to a reduction from 506 000 t to 323 000 t of production wastes in 1995 [273].

Hoechst. The year 1995 saw a total waste production of 324 000 t, comprising 72 400 t of household-type waste, 138 300 t of sewage sludge, and 113 300 t of special and hazardous wastes. Figure 127 gives a breakdown by these same three waste categories for the period 1983–1995. The marked decline in total wastes generated and also in hazardous wastes can be attributed both to the shutdown of certain production operations and to production-integrated environmental protection [274].

Waste disposal in 1995 breaks down as 132 000 t combusted and 192 000 t deposited in landfills.

5. Summary and Outlook

Integrated environmental protection can be implemented in the chemical industry by process redesign and reutilization of residues with the aim of reducing and avoiding pollutants. Because "Pollution prevention is the solution to pollution" [274a]. This can be applied as long as it is technically and economically possible. New processes for utilization of product wastes (recycling) were developed as well as the appropriate disposal of the unusable production residues as waste. These processes will be further developed following the way of innovation, i.e., technological change. Practical, scientifically based tools for material and cost management are available for this purpose. This ensures an ecologically and economically responsible utilization of scarce resources. Integrated environmental protection has two diametrically opposed aspects: ecology and economy, as shown in Figure 128. Both terms: ecology and economy contain the Greek root "οικοσ" which means house. Thus expressed differently these terms mean emphasis on "Nature's housekeeping" and emphasis on "Society's housekeeping" [275], [276].

Both concepts — their relationship cannot be conflict-free — start from the same point and are united and almost in equilibrium, as a positive contribution to the aims of sustainable development.

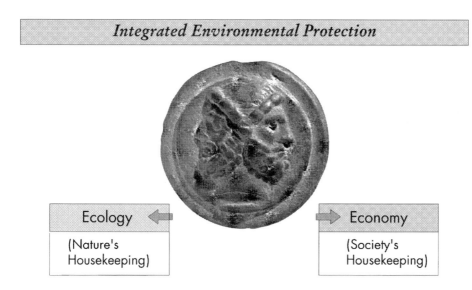

Figure 128. Integrated environmental protection

6. References

[1] Nach H. Middelhoff: "Die Organisation des betrieblichen Umweltschutzes in der schweizerischen und deutschen chemischen Industrie", *Dissertation Nr. 1293* Hochschule St. Gallen 1992.

[2] C. Christ: Umweltschonende Technologien aus Sicht eines internationalen Unternehmens — Verfahrensverbesserungen und Stoffkreisläufe, in H. Kreidebaum (Hrsg.): *Umweltmanagement in mittel- und osteuropäischen Unternehmen*, Verlag Wissenschaft & Praxis R. Brauner GmbH, Sternfels und Berlin, 1996, p. 67.

[3] World commission on Environmental and Development: *Our Common future*, University Press, Oxford, 1987.

[4] Der Rat der Sachverständigen für Umweltfragen: *Umweltgutachten 1994*, Verlag Metzler-Poeschel Stuttgart 1994.

[5] C. Christ: *Integrated environmental protection reduces environmental impact*, Chemical Technology Europe **3** (1996) p. 19.

[6] UN-Konferenz für Umwelt und Entwicklung: "Agenda 21", Rio de Janeiro 1992.

[7] M. Faber et al.: "Was ist und wie erreichen wir eine nachhaltige Entwicklung?", in U. Steger (Hrsg.): *Handbuch des integrierten Umweltmanagements*, R. Oldenburg Verlag München, Wien 1997.

[8] M. Faber et al.: "Limits and perspectives of the concept of a sustainable development", *Economie Appliquée* (1995) p. 231.

[9] R. Mannstetten: "Zukunftsfähigkeit und Zukunftswürdigkeit. Philosophische Bemerkungen zum Konzept der nachhaltigen Entwicklung", *GAIA* **5** (1996) p. 291.

[10] M. Faber, R. Mannstetten, J. Proops: *Ecological Economics — Concepts and Methods*, Edward Elgar Ltd. Cheltenham/UK, 1996.

[11] A. Moser: "Principa Ecologica: Eco-Principles as a Conceptual Framework for a New Ethics in Science and Technology", *Science and Engineering Ethics* **1** 1994, p. 241.

[12] J. Hoffmann, K. Ott, G. Scherhorn (Hrsg.): "Frankfurt-Hohenheimer guidelines for the ethical assessment of companies", *Ethische Kriterien für die Bewertung von Unternehmen*, IKO-Verlag für Interkulturelle Kommunikation Frankfurt/M. 1997.

[13] SUSTAIN Verein zur Koordination von Forschung über Nachhaltigkeit: *Endbericht: Forschungs- und Entwicklungsbedarf für den Übergang zu einer nachhaltigen Wirtschaftsweise in Österrech, (Final Report: Research and development requirements for the transition to a sustainable economy in Austria)*, Technische Universität Graz, 1994.

[14] J. L. A. Jansen: "Sustainable Development — A Challenge to Technology", Reprints of the Symposium: *Sustainable Development, Where do we stand?*, Tagungsreihe Strategien der Kreislaufwirtschaft, Institut für Verfahrenstechnik, Technische Universität Graz 1993.

[15] Verband der Chemischen Industrie e.V. (eds.): *Umwelt-Leitlinien*, Frankfurt/Main, Feb. 1988.

[16] Verband der Chemischen Industrie (VCI): *"Sustainable Development"*, Frankfurt/Main, 1995.

[17] International Council of Chemical Associations (ICCA): "Position paper on sustainable development and the chemical industry", Arlington, Virginia 1996.

[18] C. Christ: "Umweltschutz in der chemischen Industrie – Vermindern und vermeiden von Abfällen", in: R. Jünemann (Hrsg.): *Umwelt, Logistik und Verkehr*, Praxiswissen GmbH, Dortmund 1992.

[19] A. W. v. Hofmann: *Einleitung in die moderne Chemie – Nach einer Reihe von Vorträgen, gehalten am Royal College of Chemistry zu London*, Verlag Vieweg Braunschweig 1866.

[20] C. Lange: *Umweltschutz und Unternehmensplanung – Die betriebliche Anpassung an den Einsatz umweltpolitischer Instrumente (neue betriebswirtschaftliche Forschung)* **8**, Verlag Dr. Th. Gabler, Wiesbaden 1978.

[21] Bundesminister für Umwelt, Naturschutz und Reaktorsicherheit: *Umweltbericht 1990*, Bundesanzeiger no. 145 a, Köln 1990.

[22] R. Vieregge: "Integrierter Umweltschutz aus Sicht der Umweltpolitik" in H. Kreikebaum (ed.): *Integrierter Umweltschutz – Herausforderung an das Innovationsmanagement*, Verlag Dr. Th. Gabler, Wiesbaden 1990, p. 87.

[23] K. Töpfer, *Trend Z. soziale Marktwirtschaft* **39** (1989) 44.

[24] H. Strebel: "Integrierter Umweltschutz – Merkmale, Voraussetzungen, Chancen," in H. Kreikebaum (ed.): *Integrierter Umweltschutz – Herausforderung an das Innovationsmanagement*, Verlag Dr. Th. Gabler, Wiesbaden 1990, p. 3.

[25] R. Antes: "Umweltschutzinnovationen als Chancen des aktiven Umweltschutzes für Unternehmen im sozialen Wandel," *Schriftenreihe des Institutes für ökologische Wirtschaftsforschung*, no. 16, Berlin 1988.

[26] Statistisches Bundesamt: Comments on the ecological-economic studies, personal communication, July 27, 1989.

[27] Kommission der Europäischen Gemeinschaften (eds.): *Panorama der EG-Industrie*, 1990, p. 149.

[28] "The Perspective on the International Chemical Industry," *Chemical Insight* **534** (May 1994) 1.

[29] UNEP (United Nations Environment Programme) working group: "Biotechnology for Cleaner Production", Proceedings of the first Meeting, Delft 1993.

[30] Verband der Chemischen Industrie e.V. (eds.): *Umweltbericht 1988/89*, Frankfurt/Main 1989, p. 29.

[31] J. Wiesner: *Produktionsintegrierter Umweltschutz in der Chemischen Industrie*, SATW, Zürich, VDI-GVC, Düsseldorf, DECHEMA, Frankfurt/Main 1990.

[32] K. Holoubek, J. Geywitz: "Die Integration des Umweltschutzes in das operative Betriebsgeschehen," in U. Steger (ed.): "Umwelt-Auditing – Ein neues Instrument der Risikovorsorge," *Frankfurter Allgemeine Zeitung*, Verlag Bereich Wirtschaftsbücher, Frankfurt/Main 1991, p. 97.

[33] H. Wunderlich: "Umweltschutz und Sicherheit in der Produktion," in: *Die Bayer-Umweltperspektive*, Bayer, Leverkusen 1987, p. 44.

[34] M. Faber, G. Stephan, P. Michaelis: *Umdenken in der Abfallwirtschaft – Vermeiden, Verwerten, Beseitigen*, 2nd ed., Springer Verlag, Berlin 1989.

[35] C. Christ: "Produktionsintegrierter Umweltschutz am Beispiel der Chemischen Industrie," *VDI Ber.* **899** (1991) 79.

[36] G. Scharfe, B. Seweko: "Produktionsintegrierter Umweltschutz – Das Beispiel Bayer," *Chem. Ind.*, Sonderausgabe Nordrhein-Westfalen 1991, p. 17.

[37] K.-G. Malle: "From 'End of the Pipe' Philosophy to an Integrated Approach. Examples for the Evaluation of Environmental Protection in a Chemical Company," *Kem. Kemi* **20** (1993) no. 1, 23.

[38] D. Becher: "Vermeiden, Vermindern, Verwerten – integrierter Umweltschutz in der Produktion," in: *Die Bayer-Umweltperspektive II*, Bayer, Leverkusen 1991, p. 34.

[39] K. Hansmann, *Neue Zeitschrift für Verwaltungsrecht* **9** (1990) 409.

[40] K. Hungerbühler: "Produkt- und prozessintegrierter Umweltschutz in der chemischen Industrie", *Chimia* **4a** (1995) 93.

[41] H. Hulpke, U. Müller-Eisen: "The prevent principle – Environmental protection integrated in production processes", *Environ. Sci. Pollut. Res.* **4** (1997) 146.

[42] M. Faber, F. Jöst, R. Mannstetten, G. Müller-Fürstenberger: "Joint production and environmental policy: A case study for the chlorine and sulphuric acid industry", *J. Prakt. Chem.* **338** (1996) 497.

[43] G. Müller-Fürstenberger: *Kuppelproduktion – Eine theoretische und empirische Analyse am Beispiel der chemischen Industrie"*, Physica-Verlag, Heidelberg 1995.

[44] W. Sahm: "Weitsicht statt end-of-the pipe", *Chem. Ind. (Düsseldorf)* **116** (1993) no. 7–8 14.

[45] H. Schnitzer: "Die auf Stoffanalyse basierende Implementierung von vorsorgenden integrierten Umweltschutz", *Chem.-Ing. Tech.* **70** (1998) 64.

[46] J. R. Glauber: *OPERA CHYMICA*, Verlag Thomae-Matthiae Götzens, Frankfurt/M. 1685.

[47] W. Hilger: "Chemie im Dialog: Herausforderung und Perspektiven," *VCI-Dokumentation der Veranstaltung am 4. Juli 1991 im Wissenschaftszentrum Bonn*, Frankfurt/Main, Oct. 1991.

[48] H. Meffert, M. Kirchgeorg: *Marktorientiertes Umweltschutzmanagement*, J. B. Metzler Verlag and C. E. Poeschel Verlag, Stuttgart 1992.

[49] J. Senn: *Ökologie-orientierte Unternehmensführung*, Verlag Peter Lang, Frankfurt/Main 1986.

[50] F.-H. Rohe: "Umweltschutz heute in einem Chemiekonzern—Aufgaben, Probleme, Fortschritte," in: *Die Bayer-Umweltperspektive*, Bayer, Leverkusen 1987, p. 30

[51] U. Steger: *Umweltmanagement – Erfahrungen und Instrumente einer umweltorientierten Unternehmensstrategie*, Verlag Dr. Th. Gabler, Wiesbaden 1988.

[52] U. Steger: "Integrierter Umweltschutz als Gegenstand eines Umweltmanagements," in H. Kreikebaum (ed.): *Integrierter Umweltschutz – Herausforderung an das Innovationsmanagement*, Verlag Dr. Th. Gabler, Wiesbaden 1990, p. 33.

[53] R. F. Nolte: "Neue Verfahren der Produktions- und Umwelttechnik: BMFT-Förderprogramm, Märkte und Erfolgsfaktoren, 1988," in: Der Rat von Sachverständigen für Umweltfragen (eds.): *Abfallwirtschaft-Sondergutachten 1990*, J. B. Metzler Verlag and C. E. Poeschel Verlag, Stuttgart 1991, p. 207, Tz 711.

[54] R. Coenen, S. Klein-Vielhauer, R. Meyer: *TA-Projekt* "Umwelttechnik und wirtschaftliche Entwicklung" (Project "Environmental technology and development"), *Endbericht "Integrierte Umwelttechnik – Chancen erkennen und nutzen" (Final report "Integrated environmental technology – identifying and exploiting opportunities")*, Büro für Technikfolgen-Abschätzung beim Deutschen Bundestag, TAB-Arbeitsbericht Nr. 35 (Office in the German Parliament for assessing the consequences of technology, work report no. 35); Bonn 1995.

[55] A. Steinbach, A. Werner, R. Winkenbach: "Technische Buchführung—Warum die stöchiometrische Ausbeute für die Beurteilung der Produktivität eines Verfahrens nicht geeignet ist", *Chemie-Technik* **11** (1996), 41.

[56] R. Sheldon: "Consider the environmental quotient", *ChemTech* **3** (1994), 38.

[57] A. Steinbach, R. Winkenbach: "Systematische Verfahrensentwicklung/-optimierung mit der MFA-Produktivitätsfunktion", *Marktübersicht VERFAHRENSTECHNIK '97*, (1996), 8.

[58] A. Steinbach: "Materialflußanalyse in der Chemie—Integration von Prozeßdenken, technischen Aspekten und Controllingphilosophie", *m&c Management & Computer* **2** (1994) 181.

[58a] A. Steinbach: "Betriebswirtschaftlich-Technisches Controllingsystem für den produktionsintegrierten Umweltschutz in der Chemie (BTC-System)", *AbfallwirtschaftsJournal* No. 5 (1998), p. 46.

[59] A. Steinbach: "Material- und Kostenflußanalyse zur Effizienzsteigerung in der Verfahrensentwicklung", *VDI-Bericht* **1282** (1996), 152.

[60] F. Schmidt-Bleek: *Wieviel Umwelt braucht der Mensch? MIPS das Maß für ökologisches Wirtschaften*, Birkhäuser Verlag Basel 1994.

[61] C. Liedtke, T. Orbach, H. Rohn: "Betriebliche Kosten- und Massenrechnung — Ein neuer Ansatz der ökologieorientierten Kostenrechnung", Paper of Wuppertal Institut für Klima, Umwelt, Energie GmbH, vorgesehen für Veröffentlichung in Loseblattsammlung, *"Betriebliches Umweltmanagement"*, Springer-Verlag.

[62] M. Müller, P. Hennicke: *Wohlstand durch Vermeiden — Mit der Ökologie aus der Krise*, Wissentschaftliche Buchgesellschaft Darmstadt 1994.

[63] ConAccount — Support for policy towards sustainibility by Material Flow Accounting, *Conference at Wuppertal*, September 1997.

[64] A. Adriaanse, S. Bringezú et al.: *Stoffströme: Die materielle Basis von Industriegesellschaften (Resource flows — the material basis of industrial economies)*, Birkhäuser Verlag Basel 1998.

[64a] H. H. Hulpke, M. Marsmann: "Ökobilanzen und Ökovergleiche", *Nachrichten aus Chemie, Technik und Laboratorium*, 42 no. 1 (1994) p. 11.

[64b] F. Saykowski, M. Marsmann: "Ökobilanzen Fortschrittsbericht", *Z. Umweltchem. Ökotox.* 9 (1997) p. 112.

[65] U. Ullrich: "Ökologische Produktionskonzepte – Skizzen für ein naturverträgliches Leben, Arbeiten und Wirtschaften," in E. Schmidt (ed.): "Ökologische Produktionskonzepte – Kriterien, Instrumente, Akteure," *Schriftenreihe des Instituts für ökologische Wirtschaftsforschung*, no. 23, Berlin 1989.

[66] C. Christ: "Umweltschutzmanagement in der Chemischen Industrie am Beispiel der Hoechst AG," in D. Adam (ed.): *Schriften zur Unternehmensführung Nr. 48 "Umweltschutzmanagement in der Produktion,"* Verlag Dr. Th. Gabler, Wiesbaden 1993, p. 57.

[67] C. Christ: "Produktionsintegrierte Umweltschutzmaßnahmen im Lösemittelbereich," *Chem. Tech.* (Heidelberg) 19 (1990) no. 9, 42.

[68] C. Christ: "Produktionsintegrierter Umweltschutz – Chancen und Grenzen," *Chem. Ing. Tech.* **64** (1992) no. 10, 889.

[69] M. Schneider: Umweltschutz bei Bayer, *press conference*, Leverkusen, Sep. 8, 1993.

[70] H. H. Hulpke: "Umweltschutz als gesamtunternehmerische Aufgabe: Aktives und vorbildhaftes Umsetzen," *Schweiz. Maschinenmarkt* **35** (1989) 132.

[71] H. H. Hulpke: "Naturwissenschaftliche, technische sowie wirtschaftliche Aspekte und Konsequenzen des Umweltschutzes," in: *Wirtschaft und Umwelt*, C. F.Müller Juristischer Verlag, Heidelberg 1986, p. 45.

[72] H. Kreikebaum: "Integrierter Umweltschutz durch strategische Planungs- und Controlling-Instrumente," in U. Steger (ed.): *Handbuch des Umweltmanagement*, Verlag C. H. Beck, München 1992, p. 257.

[73] H. I. Joschek: "Praxis und Perspektiven der Kreislaufwirtschaft in einem Chemieunternehmen", in DECHEMA-Jahrestagungen '98 (Tagungsband), Frankfurt am Main 1998.

[74] Bayer AG: *Umweltbericht*, Leverkusen 1993.

[75] C. Christ: "Umweltschonende industrielle Arbeitsmethoden – Verfahrensverbesserungen und Stoffkreisläufe," in: *Kreislaufwirtschaft statt Abfallwirtschaft*, Universitätsverlag, Ulm 1994, p. 123.

[76] K. Zimmermann: *Präventive Umweltpolitik und technologische Anpassung*, Internationales Institut für Umwelt und Gesellschaft am Wissenschaftszentrum Berlin für Sozialforschung, Berlin 1985.

[77] K. Zimmermann, *Konjunkturpolitik* **34** (1988) 327.

[78] K. Zimmermann: "Technologische Modernisierung der Produktion – eine Variante präventiver Umweltpolitik," in U. E. Simonis (ed.): *Präventive Umweltpolitik*, Wissenschaftszentrum Berlin

[79] K. Zimmermann: *Umweltpolitik und integrierte Technologien, der Quantität-Qualitäts-Trade-off,* Wissenschaftszentrum Berlin für Sozialforschung, Berlin 1989.

für Sozialforschung, Forschungsschwerpunkte Technik, Arbeit Umwelt, Campus Verlag, Frankfurt/Main 1988, p. 205.

[80] W. Hopfenbeck: *Umweltorientiertes Management und Marketing,* Verlag Moderne Industrie, Landsberg/Lech 1990.

[81] H. Kreikebaum: "Innovationsmanagement bei aktivem Umweltschutz in der Chemischen Industrie – Bericht aus dem Forschungsprojekt," in G. R. Wagner (ed.): *Unternehmung und ökologische Umwelt,* Verlag Franz Vahlen, München 1990, p. 113.

[82] U. E. Simonis: "Ökologische Modernisierung der Wirtschaft – Option und Restriktionen," in G. R. Wagner (ed.): *Unternehmung und ökologische Umwelt,* Verlag Franz Vahlen, München 1990, p. 29.

[83] ifo Institut für Wirtschaftsforschung: *ifo – Spiegel der Wirtschaft,* München 1990.

[84] E. Feess: *Umweltökonomie und Umweltpolitik,* Verlag F. Vahlen München 1995.

[85] M. Zlokarnik: "Produktionsintegrierter Umweltschutz in der chemischen Industrie," in: *Produktions- und produktintegrierter sowie additiver und sanierender Umweltschutz,* **2,** Springer Verlag, Heidelberg (in press).

[86] D. Becher: "Vermeiden – Vermindern – Verwerten, Integrierter Umweltschutz in der Produktion," in: *Die Bayer-Umweltperspektive II,* Bayer, Leverkusen 1991.

[87] C. Christ: "Vermeidung von Abfällen in der chemischen Industrie," *Entsorgungspraxis Spezial* **1** (1990) 26.

[88] *Ullmann,* 4th ed., **6,** 155 ff.

[89] *Winnacker-Küchler,* 4th ed., **6,** 713 ff.

[90] M. Schmitt: "Zellwolleabgase als Rohstoff zur Schwefelsäureherstellung," *Chem. Ind.* (Düsseldorf) 104 (1981) no. 5, 286 ff.

[91] R. Peichl: "Das Kelheimer Modell – ein neuartiges Recycling-Verfahren," *Chem. Tech.* (Heidelberg) 11 (1982) no. 1, 21 ff.

[92] J. Dörges, K. Dietrich, B. Sülzer, J. Wiesner: "Schwefelsäure aus den Abgasen der Zellwolleproduktion," in: *Produktionsintegrierter Umweltschutz in der chemischen Industrie,* DECHEMA, Frankfurt/Main 1990.

[93] Les Usines de Melle, US 2 936 321, 1960 (J. Mercier).

[94] C. Christ: "Umweltschonende industrielle Arbeitsmethoden – Verfahrensverbesserungen und Stoffkreisläufe," in Bundesverband der Deutschen Industrie e.V. (eds.): *Industrieller Umweltschutz,* Tagungsband zum Seminar in Zusammenarbeit mit der Universität Leipzig vom 5. bis 10. April 1992 in Leipzig, Köln, Mai 1993, p. 210.

[95] Hoechst, DE-OS 36 11 886, 1987 (K. Schuchardt, H. Scholz, H. W. Neuß, H. Adam).

[96] E. I. du Pont de Nemours, GB 2 040 897, 1980 (L. J. Harwell).

[97] Knapsack, GB 1 070 515, 1967 (A. Czekay, A. Jacobowsky, E. Kottenmeier).

[98] Hoechst, US 3 980 758, 1976 (R. Krumböck, W. Kühn).

[99] ICI Australia, US 4 464 351, 1984 (V. Vasak, J. Sencar, K. Mok).

[100] Hoechst, EP-A 0 362 666, 1990 (W. Freyer, T. Olffers).

[101] Hoechst, EP-A 0 464 758, 1992;DE-OS 4 021 408, 1992 (W. Freyer, T. Olffers, A. Riedel).

[102] C. Christ: "Integrierter Umweltschutz – Strategie zur Abfallminderung und -vermeidung," *Chem. Ing. Tech.* **64** (1992) 431.

[103] P. Lappe, H. Springer, J. Weber: "Neopentylglykol als aktuelle Schlüsselsubstanz," *Chem. Z.* **113** (1989) 293.

[104] Hoechst, EP 0 278 016 B 1, 1989 (N. Breitkopf et al.).

[105] P. Sckuhr: "Umweltschutz eine ständige Herausforderung in der Verfahrensentwicklung," in: *125 Jahre Hoechst AG – Wissenschaftliches Symposium,* May 19–20, 1988, Frankfurt/Main.

[106] K. Matsumoto: "Production of 6-APA, 7-ACA, and 7-ADCA by Immobilized Penicillin and Cephalosporin Amidase," in A. Tanaka (ed.): *Industrial Application of Immobilized Biocatalysts,* Marcel Dekker, New York 1993, p. 67.

[107] U. Giesecke, F. Wedekind, W. Tischer: "Biocatalytic 7-Aminocephalosporanic Acid Production," in: *Dechema Biotechnology Conferences* **5,** Part B (1992) 609.

[108] Hoechst, DE 4 136 389, 1991 (W. Aretz, K. Koller, G. Rieß).

[109] Hoechst AG, Europäische Patentanmeldung Nr. (European patent application no.) EP 0646637, 1994, inventors: H. Bahrmann et al.

[110] Hoechst AG, Europäische Patentanmeldung Nr. (European patent application no.) EP 0646563, 1994, inventors: H. Bahrmann et al.

[111] Hoechst AG, Europäische Patentanmeldung Nr. (European patent application no.) EP 0562451, 1993, inventors: H. Bahrmann et al.

[112] Hoechst AG, Europäische Patentanmeldung Nr. (European patent application no.) EP 0597377 A1, inventors: G. Korb et al.

[113] Hoechst AG, United States Patent No. 5, 475, 107, inventors: G. Korb et al.

[114] C. Christ: "Integrierter Umweltschutz, Strategie zur Abrfallverminderung und -vermeidung" in: Tagungsband von Jahrestagung SATW 1991: *"Symposium Technik versorgen – Technik entsorgen",* Schweizerische Akademie der Technischen Wissenschaften (SATW), Zürich 1991.

[115] C. Christ: "Integrierter Umweltschutz – Strategie zur Abfallverminderung und -vermeidung in der Hoechst AG", in G. Burgbacher, K. Roth (Hrsg.): *Konzepte der Abfallwirtschaft,* **2,** Expert-Verlag, Ehningen 1992.

[116] *Houben-Weyl,* 4th ed., **9,** p. 475.

[117] *Houben-Weyl,* 4th ed., **9,** p. 476.

[118] *Houben-Weyl,* 4th ed., **10/1,** p. 492.

[119] *Houben-Weyl,* 4th ed., **6/1 c,** p. 211.

[120] *Houben-Weyl,* 4th ed., **6/1 c,** p. 221.

[121] D. Becher: *Die Bayer-Umweltperspektive II,* press conference, Sep. 16–18, 1991, Leverkusen, p. 41 ff.

[122] Bayer AG: *Umweltbericht,* Leverkusen 1993, p. 46 ff.

[123] Bayer, DE-OS 2 703 076, DE-OS 2 747 714, 1977 (O. Barth et al.).

[124] Bayer, DE-OS 4 212 086, 1992 (W. Ludwig et al.).

[125] D. I. Klukas, Leuna Werke, paper presented at the Membran-Kolloquium, Aachen, March 11, 1992.

[126] *Handbuch Chlorchemie II,* Umweltbundesamt Text 42/92, pp. 101–138.

[127] Umweltbundesamt (BUA) (ed.): "Chlortoluole," Stoffbericht 38.

[128] *Ullmann,* 4th ed., **9,** 512–513.

[129] Dow Chem., US 1 946 040, 1931 (W. C. Stoesser).

[130] Hooker Chem. & Plas., US 4 024 198, 1975 (H. E. Buckholtz).

[131] Hooker Chem. & Plas., US 4 031 142, 1975 (J. C. Graham).

[132] Hooker Chem. & Plas., US 4 031 147, 1975 (J. C. Graham).

[133] Hooker Chem. & Plas., US 4 069 264, 1976 (H. C. Lin).

[134] Hooker Chem. & Plas., US 4 069 263, 1977 (H. C. Lin).

[135] Hooker Chem. & Plas., US 4 250 122, 1979 (H. C. Lin).

[136] Ihara Chem. Ind., US 4 289 916, 1980 (Y. Nakayama).

[137] Hodogaya Chem., EP 63 384, 1981 (R. Hattori).

[138] Hoechst, EP 173 222, 1984 (H. Wolfram).
[139] Atochem, EP 126 669, 1983 (R. Commandeur).
[140] Ihara Chem. Ind., JP 60 125 251, 1983 (J. Kiji).
[141] Ihara Chem. Ind., JP 60 136 576, 1983 (J. Kiji).
[142] Ihara Chem. Ind., JP 61 171 476, 1985 (J. Kiji).
[143] Bayer, US 4 851 596, 1987 (F. J. Mais).
[144] Bayer, US 4 925 994, 1988 (F. J. Mais).
[145] Bayer, US 4 990 707, 1988 (F. J. Mais).
[146] Bayer, US 5 105 036, 1990 (F. J. Mais).
[147] *Houben-Weyl,* 4th ed., **9,** p. 477.
[148] *Houben-Weyl,* 4th ed., **9,** p. 502.
[149] *Houben-Weyl,* 4th ed., **9,** p. 436.
[150] M. Bueb, M. Finzenhagen, T. Mann, K. Müller, *Korresp. Abwasser* **37** (1990) 542–558.
[151] H. Emde, G. Randau, G. Scharfe in: *Produktionsintegrierter Umweltschutz in der chemischen Industrie,* DECHEMA, Frankfurt/Main 1990, pp. 25–26.
[152] Bayer: *Bayer-Umweltperspektive II,* press conference, Leverkusen 1991, p. 11.
[153] Bayer. "Erfolge, Projekte, Ausblick," *Umweltschutz 2/91,* p. 87.
[154] D. Tegtmeyer: "Möglichkeiten und Chancen einer membrantechnischen Abwasserbehandlung in der Textilfärberei," *Melliand Textilber.* **74** (1993) no. 2, 148–151.
[155] R. Rautenbach, A. Gröschl: "Umkehrosmose organisch-wäßriger Lösungen," *F & S Filtrieren Separieren* **4** (1990) no. 4, 231–239.
[156] T. Janssen, W. Samhaber: "Industrielle Nanofiltration zur Vorbehandlung von chemischen Abwässern," *Sandoz report,* Basel 1993.
[157] P. Meyer, W. Samhaber, C. Hamelbeck: "Vorreinigung von problematischem Chemieabwasser durch Kombination von Adsorption, Membrantrennung und Oxidation," Lecture held at an *Uhde-Symposium at IFAT,* Munich, May 25, 1990.
[158] VDI-Gesellschaft Verfahrenstechnik und Chemieingenieurwesen: "Kombinierte Verbrennung von Klärschlamm und energiereichem hochchloriertem Kohlenwasserstoff, *GVC-Tagung Entsorgung von Sonderabfällen durch Verbrennung,* Baden-Baden 1989.
[159] Abfallverbrennung am Beispiel des Bayerwerkes Leverkusen, Ein Teil des umweltgerechten Gesamtkonzeptes, *Umwelt und Technik,* 6/1988.
[160] E.-H. Rohe in: *Die Bayer-Umweltperspektive II,* press conference, Leverkusen, Sep. 16–18, 1991, pp. 27–28.
[161] Inauguration of the new Wastewater and Sewage Sludge Incineration Plant, December 2, 1988, in Leverkusen, *documentation of lectures,* p. 13.15.
[162] U. Kaier, "Chancen und Grenzen der Kraft-Wärme-Kopplung," *VDI Ber.* **808,** 1990.
[163] B. Linnhoff, "Pinch Analysis, a State-of-the-Art Overview," *TransI Chem E.* **71 Part A** 9/1993.
[164] U. Neumann, *Chem. Tech. (Heidelberg)* **12** (1983) no. 7, 21–24.
[165] BASF, Umweltbericht 1997, Ludwigshafen.
[166] BASF, DE 4 326 952, 1993.
[167] BASF, DE 4 326 953, 1993.
[168] F. Hoffmann-LaRoche, EP 5746.
[169] F. Hoffmann-LaRoche, EP 5747.
[170] S. Jensen et al.: "On the Chemistry of EDC Tar and its Biological Significance in the Sea," *Proc. R. Soc. London B* **189** (1975) 333–346.
[171] Wacker, DE 3 817 938, 1988 (L. Schmidhammer et al.).
[172] BASF, US 4 233 280, 1980 (G. Devroe, J. Langens).

[173] Stauffer Chem., *Hydrocarbon Process.* **60** (1981) no. 11, 170.
[174] L. Schmidhammer: "Rohstoffrecycling bei der Vinylchloridherstellung," *NIU Chem.* **4** (1993) no. 16, 48–50.
[175] K. Olie et al., *Chemosphere* **12** (1983) 627–636.
[176] Wacker, EP 0 177 013, 1985 (G. Dummer, L. Schmidhammer, P. Hirschmann, G. Stettner).
[177] Wacker, EP 0 180 998, 1985 (G. Dummer et al.).
[178] G. Dummer, L. Schmidhammer: "Neues Verfahren zur Entfernung von aliphatischen Chlorkohlenwasserstoffen aus Abwässern mittels Adsorberharzen," *Chem. Ing. Tech.* **56** (1984) no. 3, 242–243 (Synopsis); *VDI Ber.* **545** (1983) 885–901.
[179] Degussa, Wacker, EP 0 461 431, 1991 (K. Deller, H. Krause, L. Schmidhammer, W. Dafinger).
[180] Wacker, US 4 754 088, 1988 (L. Schmidhammer, P. Hirschmann, H. Patsch, R. Straßer).
[181] Wacker, EP 0 225 617, 1986 (L. Schmidhammer et al.).
[182] Wacker, EP 0 452 909, 1991 (L. Schmidhammer, K. Haselwarter, H. Klaus, K.-P. Mohr).
[183] Wacker, EP 0 002 501, 1978 (L. Schmidhammer, H. Frey).
[184] Wacker, EP 0 052 271, 1981 (G. Dummer, L. Schmidhammer, R. Straßer).
[185] Wacker, DE 2 803 285, 1978 (L. Schmidhammer, E. Selbertinger).
[186] Wacker, EP 0 180 995, 1985 (L. Schmidhammer et al.).
[187] Wacker, EP 0 180 925, 1985 (L. Schmidhammer et al.).
[188] G. Dummer, L. Schmidhammer: "Großtechnische Nutzung der Thermokompression bei der Destillation von 1,2-Dichlorethan," *Chem. Ing. Tech.* **63** (1991) no. 1, 74–75 (Synopsis 1914).
[189] G. Lohrengel, *Gewässerschutz Wasser Abwasser* **135** (1993) 249.
[190] G. Buxbaum (ed.): *Industrial Inorganic Pigments*, VCH Verlagsgesellschaft, Weinheim 1993, pp. 43–71.
[191] T. Fryer, *Processing (Sutton, Engl.)* (1989) July, 17–21.
[192] U. Rothe, K. Velleman, H. Wagner, *Congr. FATIPEC* **21** (1992) no. 2, 167–186.
[193] F. Wagener, J. Wiesner in: *Produktionsintegrierter Umweltschutz in der chemischen Industrie*, DECHEMA, Frankfurt/Main 1990, pp. 91–95.
[194] K.-G. Malle, *Chem. unserer Zeit* **21** (1987) no. 1, 9–16.
[195] U. Rothe in: *Sulphur 88*, Preprints of British Sulphur's 13th International Conference, Wien 1988, pp. 87–95.
[196] D. Hody, P. Bansal, *Sulzer Tech. Rev.* **69** (1987) no. 1, 25–28.
[197] H. Thumm in: *Recycling International, 4th International Recycling Congress*, EF-Verlag, Berlin 1984, pp. 1074–1077.
[198] G. Dieckelmann, H. J. Heinz: *The Basics of Industrial Oleochemistry*, P. Pomp, Essen 1989.
[199] H. J. Richtler, J. Knaut, *Fat. Sci. Technol.* **93** (1991) no. 1, 1–13.
[200] H. Pardun, *Fette Seifen Anstrichm.* **85** (1983) no. 1, 1–18.
[201] J. Baltes: *Gewinnung und Verarbeitung von Nahrungsfetten*, Parey, Berlin 1975.
[202] A. Davidsohn, *Seifen Öle Fette Wachse* **114** (1988) no. 15, 595–600.
[203] A. Zellner: Dissertation, Universität GH Duisburg 1989.
[204] K. Yoshiharu, O. Toshio, US 4 164 506, 1978.
[205] H. Druckrey et al., *Z. Krebsforsch.* **74** (1970) 241–270.
[206] Henkel, DE-OS 3 501 761, 1986 (L. Jeromin, E. Peukert, G. Wollmann).
[207] Y. M. Choo, K. Y. Cheah, A. N. Ma, A. Halim: "Conversion of Crude Palm Kernel Oil into its Methyl," *Proc. World Conference on Oleochemicals* 1990 (Publ. 1991, pp. 292–295).
[208] B. Gutsche, V. Jordan, L. Jeromin: *Chem. Ing. Tech.* **64** (1992) 774.
[209] B. Freedman, E. H. Pryde, T. L. Mounts, *JAOCS J. Am. Oil Chem. Soc.* **61** (1984) no. 10, 1638–1643.

[210] Henkel, DE-OS 3 932 514, 1991 (G. Aßmann et al.).
[211] Henkel, DE-OS 3 911 538, 1991 (G. Aßmann et al.).
[212] H. Wiederkehr: "Reduktion der Umweltbelastung durch Prozeßoptimierung in der chemischen Produktion," *Chimia* **40** (1986) no. 9, 323–330.
[213] F. Hoffmann-La Roche, CH 642 936, 1984.
[214] J. Jeisy: Vortrag auf dem Jahrestreffen der Verfahrens-Ingenieure, Sep. 29, 1982, Basel.
[215] J. Jeisy et al.: "Beseitigung acetylenhaltiger Abgase," *Chem.-Ing.-Tech.* **55** (1983) no. 5, 388–389.
[216] G. H. Lenske, J. Findling: "Ein Gasstrahler zur Förderung von Kopfgasen der Rohöldestillation ins Heizgasnetz," *Erdöl Kohle Erdgas Petrochem.* **35** (1982) no. 2.
[217] H.-G. Franck, J. W. Stadelhofer: *Industrielle Aromatenchemie*, Springer Verlag, Berlin 1987, p. 36.
[218] H.-G. Franck, J. W. Stadelhofer: *Industrielle Aromatenchemie*, Springer Verlag, Berlin 1987, pp. 324–344.
[219] S. Rittner, R. Steiner, *Chem. Ing. Tech.* **57** (1985) no. 2, 91–102.
[220] A. Papp, K. Saxer, *vt Verfahrenstech.* **15** (1981) no. 3.
[221] Shell, US 3 562 239, 1966.
[222] Shell, US 4 414 132 and 4 329 253, 1979.
[223] *The Clean Air Act and the Refining Industry*, UOP, Sep. 1991.
[224] C. Higman, G. Grünfelder: "Clean Power Generation from Heavy Residues," *Proc Inst Mech Eng, IMech E Conf* 1990, no. 13.
[225] P. Ladeur, H. Bijwaard: "Shell plans $ 2.2-billion Renovation of Dutch Refinery," *Oil Gas J.* **91** (1993) no. 17, 64–67.
[226] E. Stöldt: "Concepts for Zero Residue Refineries," *Interpec China '91*, Beijing, Sep. 1991.
[227] H. Heurich, C. Higman: "Partial Oxidation in the Refinery Hydrogen Management Scheme," *AIChE Spring Meeting*, Houston, 1993.
[227a] C. Higman, M. Eppinger: "The Zero-Residue Refinery using the Shell Gasification Process," *Achema 1994*, Frankfurt/Main.
[228] M. J. Niksa: "Acid/Base Recovery from Sodium Sulphate," *5th International Forum on Electrolysis in the Chemical Industry*, Electrosynthesis Inc., Nov. 10–15, 1991.
[229] H. V. Plessen et al., *Chem. Ing. Tech.* **61** (1989) no. 12, 933–940.
[230] Olin, US 5 084 148, 1992.
[231] Hoechst, US 4 613 416, 1986.
[232] J. Jorisse, K. H. Simmrock, *J. Appl. Electrochem.* **21** (1991) 869–876.
[233] E. Hillrichs, K. Lohrberg, *Dechema Monogr.* **125** (1992) 221–232.
[234] General Electric, US 4 561 945, 1985.
[235] De Nora Permelec, WO 93/0046, 1993.
[236] U.S. Dept. of Energy, US 4 876 115, 1989.
[237] Chem. Systems Inc.: Bipolar Membranes, *Report* no. 86-9, USA, Dec. 1987.
[238] K. N. Mani et al., *Desalination* **68** (1988) 149–166.
[239] B. Pieper: "Natriumsulfat-Elektrolyse in der Membranzelle," *Reihe 3*, VDI Verlag, Düsseldorf 1987.
[240] G. Cowley et al.: "Electrochemical Generation of Chloric Acid at High Current Efficiencies," *5th International Forum on Electrolysis in the Chemical Industry*, Electrosynthesis Inc., Nov. 10–15, 1991.
[241] Bayer, US 5 071 522, 1991.
[242] N. Kimura: *Senjo-Sekkei*, Kindai-Hensyu Ltd., Tokyo 1991, pp. 8–12.

[243] H. Matsuzaki: *Senjo-Sekkei*, Kindai-Hensyu Ltd., Tokyo 1987.

[244] T. Ohmi et al.: Nikkei Microdevices, May 1988, p. 101.

[245] N. Miyake, Y. Matsuo: Outline of Membrane Operation Technology, Fuji Techno System Ltd., Tokyo 1991, p. 933.

[246] Y. Sugimoto in, 245 p. 683.

[247] Y. Hiratsuka, H. Sato, N. Hashimoto, T. Shinoda: *Senjo-Sekkei*, Kindai-Hensyu Ltd., Tokyo 1989, pp. 2–12.

[248] M. Okada, J. Yonamoto, *Ultraclean Technology* **3** (1991) no. 2, 203.

[249] Boehringer Mannheim (ed.): *Umweltschutz durch Biotechnik*, Mannheim 1990.

[250] Statistisches Bundesamt: *Umwelt*, Fachserie 19, Reihe 1.2: "Abfallbeseitigung im produzierenden Gewerbe und in Krankenhäusern," Wiesbaden 1993.

[251] Statistisches Bundesamt: personal communication.

[252] H. Hulpke, U. Müller-Eisen: "Abfallmanagement in der Chemischen Industrie," *Chem. Umwelt Tech.* 1993, 22.

[253] C. Christ: "Konzepte zur Abfallverminderung und -vermeidung in der Chemischen Industrie," in H. Sutter, M. Held: *Stoffökologische Perspektiven der Abfallwirtschaft*, E. Schmidt Verlag, Berlin 1993, p. 83.

[254] W. Hilger: *Chemie im Dialog "Herausforderung und Perspektiven,"* VCI-Dokumentation der Veranstaltung am 04.07.1991 im Wissenschaftszentrum Bonn, Frankfurt/Main, Oct. 1991.

[255] H.-I. Joschek, G. Janisch: "Entsorgungskonzept eines Großbetriebes am Beispiel der BASF AG Ludwigshafen," in K. J. Thomé-Kozmiensky: *Sonderabfallwirtschaft*, EF-Verlag für Energie- und Umwelttechnik, Berlin 1993.

[256] BASF, *Umweltbericht 1997*, Ludwigshafen.

[257] H.-I. Joscheck, I. H. Dorn, T. Kolb: "Der Drehrohrofen. Die Chronik einer modernen Technik am Beispiel der BASF-Rückstandsverbrennung," *VGB Kraftwerkstechnik* **75** (1995) no. 4.

[258] R. Römer, N. Neuwirth: "Zentrale Schwermetallfällanlage der BASF AG für Abwässer aus Rauchgaswäschen. Verfahrenstechnik der mechanischen thermischen, chemischen und biologischen Abwasserreinigung", *GVC-Symposium*, Baden-Baden 1988.

[259] K. Holzer: "Aerosolabscheidung aus Rauchgasen von Anlagen zur Verbrennung von Sonderabfällen," *Staub Reinhalt. Luft* **48** (1988) 203.

[260] K. Capek: "Gemeinsame Verbrennung von Industrie-Klärschlamm und flüssigem Sonderabfall," *VGB Kraftwerkstech.* **70** (1990) 31.

[261] Hoechst DE-OS 3 703 568 A 1, 1988 (G. Müller, H. Thomas, H. Schmolt).

[262] B. Koglin, J. Beyer, R. Rink, J.-E. Roth: "Hochdruck-Verdichtung und Einbau von Abfällen, ein Systembestandteil der Bayer Kompakt-Deponie," *Chem. Ing. Tech.* **63** (1991) 605.

[263] W. Regenberg: "Design and operation of BASF's landfill site and underground storage for industrial waste", *5th International Symposium on Operating European Hazardous Waste Treatment Facilities*, Odense, September 1992.

[264] "Multimineralische CONTREB-Kombinationsdichtung, Deponie Flotzgrün", Bilfinger + Berger AG, Mannheim; BASF AG, Ludwigshafen 1993.

[265] Kali-Chemie, EP-A 0 474 093 A 1, 1991 (W. Legat et al.).

[266] Kali-Chemie, EP-A 0 545 851 A 1, 1992 (H.-W. Swidersky et al.).

[267] Kali-Chemie, EP-A 0 546 984 A 1, 1992 (H.-W. Swidersky et al.).

[268] BASF, Pressegespräch zum rohstofflichen und werkstofflichen Recycling von Altkunststoffen bei der BASF, Ludwigshafen, March 16, 1994.

[269] Hoechst, EP 0 212 410 B 1, 1990 (S. Hug et al.).

[270] Verband der Chemischen Industrie e. V.: Pilotprojekt, Verwertung gebrauchter Industriepackmittel aus der Chemischen Industrie, Frankfurt/Main 1991.

[271] H.-I. Joscheck: "Praxis und Perspektiven der Kreislaufwirtschaft in einem Chemieunternehmen", in: *DECHEMA-Jahrestagungen '98 (Tagungsband)*, **II,** Frankfurt a. M. 1998, p. 65.

[272] H.-I. Joscheck, G. Janisch: "Vermeiden, Vermindern, Verwerten und Beseitigen — das Entsorgungskonzept der BASF AG, Ludwigshafen", *UmweltWirtschaftsForum* **4** (1996) 67.

[273] Bayer, *Umweltbericht 1997*, Leverkusen.

[274] Hoechst, *Umweltschutz-Jahresbericht 1995*, Frankfurt/Main.

[274a] D. Huisingh: "Cleaner Production: Concepts, Processes and Procedures that really work", *Chemistry for sustainable development* 1 (1993) p. 83.

[275] M. Faber, R. Mannstetten, J. Proops: "Toward an open future: Ignorance, novelty, and evolution", in B. Constanza, B. Norton, B. D. Haskell (eds.): *Ecosystem Health, New Goals for Environmental Management,* Washington 1997, p. 72.

[276] P. Ehrlich; "The limits to substitution: Meta-resource Depletion and a New Economic-Ecological Paradigm", *Ecological Economics* **1** (1989) p. 9.

Index

α-glucosidase   158
1-Amino-8-hydroxynaphthalene-3,6-disulfonic acid   67
1-butene   56
2-acetaminonaphthalene-8-sulfonic acid   42
2-butene   56
2-naphthylamine-8-sulfonic acid   40
3-methylxanthine   57
7-Aminocephalosporanic acid   53
acetylation   40
acetylene   107
acetylene process   108, 110
acid sodium sulfate   145
additive environmental protection   9, 14, 15, 16
adsorber plant   102
Agenda 21   1
alkaline refining   116
alkanesulfonates   71
amyl alcohol   56, 57
approaches to environmental protection   14
asbestos disposal   174
associated substances   7
atom utilization   22
auxiliaries   8
azeotropic mixture   40

basic balance sheet   24
benzine recovery   64
benzothiazepine   74
biocatalysis   156
biocatalytic splitting   156
brominated organics   43
Brundtland Report   1
by-products   8

caffeine   57
capital investment   12, 31
carbon disulfide recovery   38, 39
carbon slurry   139
catalyst change   74
catalysts   8
caustic soda concentration   148
chemical industry wastes   163
chemical refining   129
chloride process   113
chlorinated hydrocarbon side products   81
chlorinated hydrocarbons   82
chlorination of toluene   73
chlorination processes   47
chlorination residues   49
chlorine recovery rate   99
chromium circulation   54
clean production   11
clean technology   11
closed production cycles   45

closed recycling of auxiliaries   75
cobalt catalyst   56
cocatalysts   74
combustion unit   168
concepts of environmental protection   12
continous oxidation column   54
conventional sulfate process   111
Cost Flow Analysis   12, 21, 26
costs   31
cumene   104
cumene production   105

DENOX catalysts   91
diluent – slurry process   133
disposal measures   167
disposal of TPPO   90
distillation unit   153
distillation without energy transfer   66
downcycling   167
dry acetylation   42
dye production   79

e. coli   55, 160
ecological sustainability   2
economic limitations   29
economic problems   2
effect of production-integrated environmental protection   29
electrolyzers   143
elementary cells   143
emission reduction   87
emissions into the atmosphere   16, 18
end of pipe measures   9
end-of-pipe technologies   21, 165
energetically coupled distillation   67
energy conversion   88
environmental concepts   9
environmental protection (EP)   9
enzymatic process   55, 56
EP related to production   9
ester waxoil production   51
etinol   122, 124
etinol production   123, 125
exhaust gases   47

factors of sustainability   2
fatty acid methyl esters   115
filtration unit   153
five-stage crystallization   130

gasification process   136
genetic engineering   55, 158, 159, 160
glucose-6-phosphate dehydrogenase   158
grid diagram   87

197

halogenated aromatics   43
hazardous waste incinerator   169, 170
hazardous wastes   168
HCl recovery   96
HCl-generating plant   98
hexanol recovery   62
holistic marketing concept   167
hydrina membrane electrolyzer   145, 146
hydrina-technology   142
hydrochloric acid   47
hydroformylation   56
hydrogen halides   43
hydrogen sulfide recovery   38
hydrogen sulfide utilization   38

impurities   7
incomplete conversion   7
industrial management   10
industrial power plants   84
integrated environmental protection   10, 15, 21, 31
integrated vinylchloride production   97
internal rate of return   29
internal recycling   50
investment decisions   29
isopropanol recycling system   151

joint products   7

landfill disposal   173
liquid propylene process   132
liquid residues   47
liquid ring pump   46
low-pressure transesterification   116, 118, 120, 121

mass flow   128
material flow accounting   27
material flow analysis   12, 21, 22
material planning productivity   21, 22, 25, 26
membrane technology   152, 71, 79
methylation   57
methylene chloride recovery   53
microorganisms   159

NIPS system   27
multiple crystallization   128
n-valeraldehyde   56, 56
nanofiltration   71, 72
naphthalene   128
naphthalene refining   129
naphthalene sulfonation   76
naphthalenedisulfonic acid   75, 78
neopentyl glycol   50
net present value   29
neutral salt splitting   142
nickel catalyst   56, 69

oleochemistry   115

operating costs   16, 17
operating costs in relation to added value   20
optimized cogeneration   84
organic solvents   59
organohalogen compounds   41
ortho-chlorotoluene   73
oxychlorination   96

paint recycling   178
para-chlorotoluene   73
partial oxidation processes   139
pay back period   29
penicillin   156
pervaporation   152
pervaporation module   154
pervaporation process   152
phase transformation catalyst   59
physical refining countercurrent crystallization   130
physical refining process   129
plant specific environmental protection   9
plasma arc process   107
plastics recycling   174
pollution control balance   39
poly (ethylene terephthalate)   28
poly (vinyl chloride)   107, 93
polypropylene   132, 175
powdered dye   80
pre-esterification   121
pressure permeation plant   80
prevention of residues   13
primary by-products   26
primary raw materials   22, 24, 25
principles of chemical production processes   165
process rearrangement   38, 79
product wastes   163, 166, 174
product-integrated environmental protection   15
product-oriented management   166
production of pure hydrochloric acid   49
production wastes   163
production-integrated environmental protection   9, 10, 12, 178, 29
production-oriented management   165
productivity function   23, 24

reaction medium   8
real balance yield   23
reclamation   166
recombinant organisms   160
recorcin production   62
recovery of acetic acid   40
recovery of alkanesulfonates   71
recovery of auxiliary products   51
recovery of benzine   62
recovery of bromine   43
Recovery of diisopropylether   61
recovery of dimethyl formamide   64
recovery of ethanol   64

recovery of hexanol   61
recovery of hydrogen sulfide   38
recovery of hydroxypivalaldehyde   52
recovery of methanol   40, 63
recovery of organic solvents   59
recovery of sodium sulfate   38
recovery of solvents   59, 61
recovery of toluene   61, 63
recovery of trichlorethylene   64
recycling of PET wastes   176
recycling of polyoxymethylene   175
recycling of PVC processing wastes   175
recycling of used plastics   176
refinery flow sheet   135
refrigerant recycling   176
refrigeration process   127
regeneration of cellulose fibers   36
residual contamination of wastewater   16, 19
residue management   165
residue reduction   166
residues   5
residues from production   164
residues in chemical processes   6
responsible care   5
revised H acid process   69
rhodium catalyst   57

secondary by-products   26
secondary raw materials   23, 24, 25
sewage sludge   17, 82
sewage sludge combustion   81
shell gasification process   135
sliding vane rotary pump   46
sludge incineration   83
sodium sulfate concentration   148
sodium sulfate electrolysis   143, 144
sodium sulfate recovery   38
sodium sulfate solutions   150
solvent recovery   59, 61
soot ash removal unit   140
specific balance yield   23
splitting of penicillin   157

stoichiometric yield   21
sulfate process   111, 112, 113
sulfuric acid   146
sulfuric acid concentration   150
sunk-cost   29
sustainability   2
sustainable development   1, 5, 12

technical limitations   28
theobromine   57, 58
theophylline   57
theoretical balance yield   22
titanium dioxide   111
titanium dioxide pigment   113
toluene recovery plant   63
toxic waste gases   35
transesterification   117
treatment process of TPPO   92
two-stage thermal treatment   49

used packaging materials   177
utilization of residues   13, 14

vinyl chloride   93
vinyl chloride monomer   94
vinylchloride production   102
viscose staple fiber production   34, 37
vitamins   122

waste avoidance   166
waste combustion   168
waste management   163, 165, 178
waste-management concepts   164
waste-gas capture   37
waste-gas emission   37
wastes from production   164
wine yeast   16
Wittig reaction   90

zero-residue refinery   135
Ziegler–Natta catalysts   135